*Methods in
Experimental Embryology
of the Mouse*

KEEN A. RAFFERTY, JR.

Methods in Experimental Embryology of the Mouse

THE JOHNS HOPKINS PRESS

BALTIMORE AND LONDON

The Johns Hopkins Press, Baltimore, Maryland 21218
The Johns Hopkins Press Ltd., London

Library of Congress Catalog Card Number 70-101642
Standard Book Number 8018-1129-5

To my wife, Nancy

Preface

Coursework dealing with the experimental embryology of mammals is rare in universities and schools of medicine. As far as I know no attempt has been made to assemble a detailed, reasonably comprehensive compendium of methods intended for this purpose, or for workers in the field and those contemplating such work. This hiatus undoubtedly results from the fact that in the past this area has been expensive and technically difficult, if not limited in application. In recent years, however, great strides have been made with respect to the streamlining and rationalization of method, to the development of new and easily applicable techniques, and to a clearer understanding of the areas in which important progress is likely to be made in the future. I believe, in fact, that advances in this area have caught many biologists unaware, and that we may well be on the verge of great progress in the study of development because of this latter-day accessibility of the mammal to experimental analysis.

The result of these developments, combined with the fortunate proximity of Dr. John Biggers' laboratory for reproductive biology, convinced me that many of the more promising methods for the experimental study of mammalian development can be carried out by medical and graduate students, and by advanced undergraduates.

Recently this point was tested when Drs. Samuel Stern and Roger Donahue joined me in offering an eight-week course for a small number of first-year medical students. In the course, many of the advanced procedures currently in use in this area were given as laboratory exercises and later utilized to conduct brief original studies. The success of the venture generated this manual. If it fares well, it will be by helping students and professional workers to enter the field.

For those who may contemplate doing so, a special feature should be emphasized. One must realize that many of the traditional experiments upon avian and amphibian embryos remain difficult to accomplish with mammals and will probably continue to do so to any degree permitting casual student laboratory work; for evident reasons, approaches to extirpation, transplantation, and regeneration studies are somewhat limited. On the other hand, the special value of mammalian development is that important embryological aspects which may be unique to this group are distinctly accessible to analysis. Here I refer to the events of early development in particular, notably cleavage, blastocyst formation, implantation, and pre-streak determination, all of which are well suited to study. There is good reason to believe that differentiation of some cell types, such as trophoblast, is precocious and not dependent upon the stimulus of classical morphogenetic movement. Perhaps associated with precocious differentiation, there is now good evidence that genetic de-repression, as evidenced by the appearance of new enzymatic activities, occurs early in cleavage of mammalian eggs, rather than at the time of gastrulation, as has been observed for other developing forms. These are only two aspects which need further study. The proper approach to the analysis of mammalian development, therefore, is to recognize that the mammal owes no apologies for its developmental mode, and deserves attention because of its very uniqueness, which has already contributed much to general aspects of developmental analysis.

Among mammals, the mouse has been singled out for exclusive attention for the related reasons that the animal is easily the most economical and has been the object of far more attention than any other mammal. Limiting the book to the mouse, therefore, enlarges the opportunity for comprehensiveness and increases its utility as a teaching device.

Because many methods are described herein, there are in effect many contributors. In addition to these, Drs. Stern and Donahue, Drs. Biggers, Fernando Porturas, and David Whittingham deserve much of the credit for whatever may be of value in this book. They helped to teach me fact and method, and were indispensable in the book's genesis. Two students in particular, Mr. James Schreiber and Miss Elizabeth Lancaster, took considerable pains to fill in gaps in published descriptions of procedures and to suggest additional material while busy at their own projects. Finally and principally, Dr. Nancy Rafferty, my wife, gave much of her time to the illustrations and helped greatly with critical suggestions. I am grateful to all.

ix

Contents

List of Illustrations

*Methods in
Experimental Embryology
of the Mouse*

Part I

Instruments, Equipment, and Materials

By far the most important piece of equipment for the experimentalist is a dissecting microscope of good optical quality which is also versatile and convenient to use. Several are available, such as the Wild M5 Stereomicroscope. Whatever microscope is used, illumination should be rapidly changeable from transmitted to incident, and a permanently mounted objective housing giving quickly changed magnifications ranging between about 6 and 50 or 100 diameters is important. The operator will find handrests necessary to provide support during manipulation. If these are not provided on the instrument itself, small wooden boxes placed alongside the microscope are satisfactory.

Since much of the work involves culturing cleaving eggs for several days, aseptic technique is important. Although not absolutely essential, a clean-area hood is very desirable, preferably one made of plexiglass or else self-illuminated. The type used in the laboratory of Dr. John Biggers, and illustrated in Figure 1, is both convenient and effective. Section drawings and instructions for building this unit are given in the appendix, but almost any covered area which accommodates the dissecting microscope and gives a reasonable working space is suitable. If no hood of any kind is available, culturing can be done with surprising success on open benches, especially if air movement is avoided.

An incubator set at 37° C. is essential for culturing experiments. In addition, since the media in use include bicarbonate, the atmosphere over the cultures should consist of 5% CO_2 in air in order to maintain the pH. The most convenient (and also the most expensive) way of accomplishing this is by means of a continuous flow gassed incubator. Alternatively, equally good results are obtained by placing cultures in small containers such as dessicating jars, replacing the atmosphere with the 5% CO_2 mixture, and incubating the jar.

A burner, preferably with a pilot flame, such as the Touchamatic, is required for flaming sterile glassware and for making micropipettes. A conventional pilotless Bunsen burner is satisfactory but less convenient. If such a burner is used, each student must make his own microburner for pulling pipettes (Fig. 2). A suitable microburner

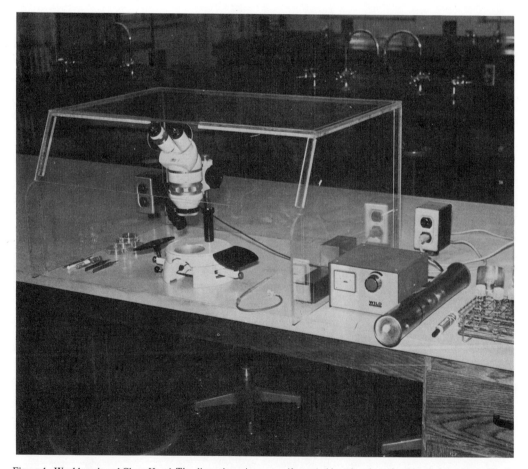

Figure 1. Workbench and Clean Hood. The dissecting microscope if covered by a hood, preferably of clear plastic. Working instruments (forceps, capillary pipette, syringe, etc.) are kept within it. Other materials, such as measuring pipettes and medium, may be kept in appropriate sterile containers outside the hood.

may be made by bending the capillary portion of a disposable Pasteur pipette; alternatively, a burner made by bending a 20-ga. syringe needle is quite satisfactory, and more permanent. Other necessary supplies and equipment include mouthpiece assemblies (such as A. H. Thomas Co., #3395M), embryological watchglasses, 9″ or 5¾″ disposable Pasteur pipettes (a large supply), two pairs of straight watchmaker forceps, several 30-ga. syringe needles, and a needle holder with a small sewing needle or its equivalent. A supply of sterile culture dishes is needed, and these may conveniently be quite small to conserve medium or oil. We find 35 mm. $\times$ 10 mm. sterile disposable tissue culture dishes (Falcon Plastics, #3001) to be suitable, but larger glass or plastic dishes may be used.

Aside from the incubator, the instruments and supplies given above are intended as the equippage for individual students or student pairs. However, there must also be access to equipment for filter sterilization of culture med-

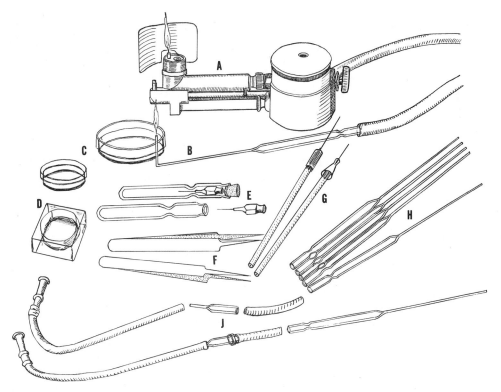

Figure 2. Instruments for Ovum Manipulation and Other Operations. A. Burner with pilot flame, also useful for pulling capillary pipettes; B. Microburner made by bending a Pasteur pipette; C. Disposable plastic culture dishes; D. Embryological watch glass; E. Altered 30-gauge syringe needles for flushing oviducts; F. Watchmaker's forceps; G. Sewing needle or small gauge syringe needle attached to holders; H. Supply of Pasteur pipettes; J. Mouthpiece assemblies, with middle adapter made by cutting the ends of a Pasteur pipette.

ium and to an autoclave for sterilization of glassware. Concerning the former, membrane, porcelain or fritted glass filters may be used. Suitable equipment is supplied by various manufacturers, such as the Millipore, Gelman, Selas, and other companies. "Swinny" sterile disposable filtration units are available from some of these manufacturers and are especially convenient for aseptic gaseous equilibration of medium and oil, as well as for sterile filtration of small fluid volumes.

General dissecting and operating instruments for removing tissues from mice are needed, but their nature will be obvious and no special listing is required. Small forceps of the tissue-gripping and of the serrated-tip variety are desirable, and wound clips and a wound clip applicator are such time-savers as to be nearly essential.

B. HUSBANDRY, MATING, AND GENERAL CONSIDERATIONS

Experience in a number of laboratories suggests that the best results, in terms of successful matings and survival of ova, result from use of partially inbred, as opposed to genetically homogeneous, mouse lines. In the United States the form most used is the random-bred, so-called Swiss, or Swiss albino mouse, such as is derived from the ICR population and sold by various dealers. Additionally, random-bred Swiss mice are considerably cheaper than those of most inbred lines, a result of their general vigor. The account that follows is tailored to behavioral and biological characteristics of this form, and it should be borne in mind that other lines may differ somewhat in various ways, especially behavioral ones. A secondary advantage of Swiss mouse stocks is the comparative docility of the animals.

In naturally mated mice, insemination occurs at about 10:00 P.M. to 12:00 P.M., and ovulation usually between 10:00 P.M. and 2:00 A.M. Fertilization is thought to occur at about the middle of the dark period, which is in the vicinity of 1:00 A.M. The gestation period is 19 to 21 days, parturition occurring between about midnight and 4:00 A.M. Females which have given birth enter estrus and mate again the next night: the postpartum estrus.

Times of events may vary in different populations, however. Data for one particular population are given in a paper by Braden and Austin (13).

The estrous cycle of the mouse is four (occasionally five or more) days, with the practical result that a maximum of one-fourth of natural matings succeed in any one night. However, regularity of the day-night cycle is quite important for promoting frequent and regular estrous cycling. Mouse rooms may be provided with a timer switch giving 14 hours of light and 10 hours of dark, or the natural day-night cycle may be used in windowed rooms. In either case, however, newly arriving animals should be acclimated for a few days before use. Additionally, it is important to avoid turning on the room lights during the night phase and to avoid undue noise or upsets in routine at any time. Humidity is probably not a significant factor except at extremes, but constancy of temperature does seem to be. A temperature of 70 to 75° F. is satisfactory, but whatever mean is used, fluctuations of more than 2° should be avoided. Strong air currents are also undesirable.

For mating, females of 7-8 weeks of age are most often used, or ones that have littered once (uniparous, or "proven" females). In practice, good results are often obtained by use of recently shipped 8-week virgin females. Males should be 12 weeks of age or older, up to 18 months. When mating failures occur the blame is usually laid to interruption of the estrous cycle in females, but male mice are also quite susceptible to upset, which may result in their refusing to mate. Immature males and males that are mature but have never mated may be kept together in stock cages. Males that have once mated, however, fight on contact, and mating does not occur if more than one male is present. Escaped males running loose in the animal room climb about on the cage tops and their presence there also seems to inhibit mating of caged pairs. Escaped males should be captured whenever located. In some animal rooms, poisoned bait is set out constantly to eliminate such animals. Since there is some evidence that the urine of a strange male inhibits male breeders (110), males are usually kept in the same individual cage and females transferred to it when matings are desired. Because males vary as effective breeders, it is worthwhile keeping a record of breeding success on each male's cage. It is customary to rest males for three days between matings, but in natural matings a male will impregnate about one female a day if a constant population of four or more females is maintained in the cage.

For timed matings, two to four females are added to

each male's cage late in the afternoon. After insemination the ejaculum coagulates, ideally resulting in a vaginal plug which is visible the next morning and permits recognition of successfully mated females. Suspect females are lifted by the tail and allowed to steady themselves by grasping the cage top with their front feet. Using a small blunt spatula, the dorsal margin of the vagina is pushed away from the pubic symphysis and the vagina inspected for a plug (Fig. 3). It should be borne in mind that the

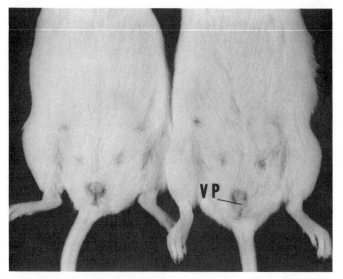

Figure 3. Female Mice with and without Vaginal Plugs. *Right:* Mouse with a vaginal plug. VP, seen just below the clitoris. *Left:* Normal control.

plugs are temporary and soon lost; inspections should be made before 10:00 A.M. on the day following mating. In some instances the plug may be small or may even consist of an inconspicuous mucous mass, but it is sometimes quite large and distinct, as shown in the figure. The author's experience has been that vaginal plugs are often present in females that do not become pregnant and are frequently absent in females that do, or that are subsequently found to contain fertilized eggs. An equally satisfactory indication of mating may be an "estrous," or expanded, vagina, though many laboratories succeed quite well in identifying mated females by careful examination for plugs. In most situations it is probably more efficient to assume that all mice have mated if hormonal stimulation is used (see below). Since some Swiss females apparently never mate, it is often also good practice to discard those that do not mate after four or five days exposure to males, or to a single night's exposure following hormonal stimulation.

In experiments where it is important to record the exact time of insemination there is no recourse but to watch pairs continuously in dim light, from the introduction of females into males' cages at about the mid-dark period until unmated females are removed several hours later. For this purpose, a red photographic safelight is helpful. The observer should not be misled by mountings by the male, since these are frequent and seldom result in coitus. Coitus is a longer act, lasting perhaps 10 to 15 seconds, and ends as the coupled pair falls to the side and rests briefly before separating. The act is distinctive from casual mounting but should be witnessed before it is necessary to make important records of coitus.

It is sometimes convenient to induce ovulation to occur at unusual times. PMS/HCG injections (p. 23) may be given between 9:00 A.M. and 10:00 P.M. to mice held on a normal day-night cycle, ovulation occurring about 12 hours thereafter. As noted below, ovulation may be obtained at other times by reversing the day-night cycle.

After coitus, coagulation of the product of the seminal vesicles occurs within minutes to form the vaginal plug. Inseminated females may be removed immediately after the event, transferred to another cage, and observed 10 or 15 minutes later to determine whether a vaginal plug has formed. Presence of a plug is confirmation of coitus; its observance soon after the act reduces the possibility of false negative results from loss of the plug within the next hours.

When mice are to be killed, the method universally used is that of cervical dislocation, since it combines humane considerations with convenience. The mouse is lifted by the tail and placed upon a screen surface, which it grasps firmly. With the other hand, the experimenter holds the base of the skull against the surface, using the thumb and index finger (Fig. 4). When the skull is pinioned, the tail is pulled firmly, stretching the neck and causing severance of the spinal cord and of the cervical nerves responsible for respiration, plus sufficient damage to the brain to result in unconsciousness. Cervically dislocated mice may be opened or autopsied immediately.

When necessary to identify individual mice kept to-

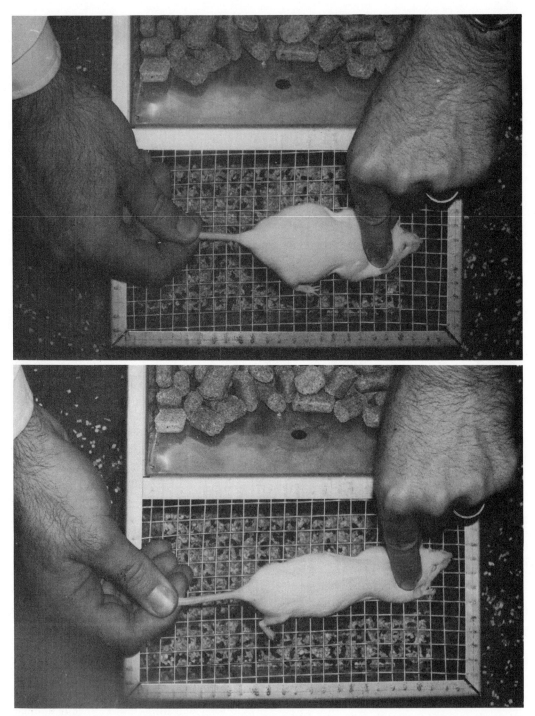

Figure 4. Cervical Dislocation: A Rapid and Humane Method of Killing Mice. *Top:* Animal is picked up by the tail and placed on the cage screen, which it grasps with its forefeet. The head is pinioned by placing thumb and forefinger behind it, approaching from behind. *Bottom:* The tail is pulled firmly, dislocating the occipital joint and causing complete paralysis. Death follows quickly from asphyxia and possibly impaired cranial circulation.

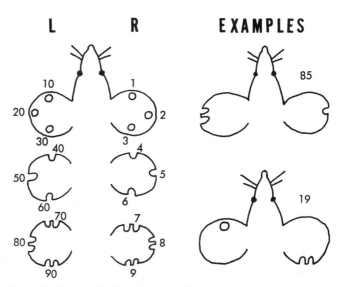

Figure 5. Code for Marking Ears. Markings are made with an ear punch. From Dickie (27).

gether in cages, ear punches are most often used, although mice may also be identified by dyes applied to the back fur. The ear punch system in use at the Jackson Laboratories in Bar Harbor, Maine, is given in Figure 5 (27).

Some work requires mice to be handled just before or after mating, which normally occurs at an inconvenient time. However, the day-night cycle may be reversed through the use of a windowless mouse room and a continuously cycling timer switch adjusted for a mid-dark period whenever wanted. The method has an added advantage in discouraging visitors (and hence noise) during daily working hours and ensures tranquility during the mouse's light period. In addition, it tends to enforce husbandry operations, such as change of bedding and water bottles, feeding, etc., into unvarying routine. All of these factors are important in maintaining proper estrous cycling. A ten-hour artificial dark period may end at 3:00 P.M. or 4:00 P.M., allowing the last two hours of the working day for maintenance operations, matings, and so forth, and providing for insemination late in the morning. Other cycles may be devised for the convenience of various kinds of experiments, but it should be remembered that mice require several days to adjust to abrupt shifts in the day-night sequence.

When surgery is to be done, the usual anesthetic is

Nembutal, which, however, is dangerous if given in a large dose to induce deep anesthesia. Perhaps the best procedure is to inject intraperitoneally a light anesthetic dose of Nembutal, and to "touch-up" when needed with ether, administered as needed via a small cone made of stiff paper and packed with a small wad of cotton at one end. The point of the cone should be cut off to facilitate ventilation.

We have had excellent results by diluting standard veterinary Nembutal (50 mg./ml.) to 10% with 0.85% saline and injecting 0.40 ml. per 30 gm. mouse.

For further details concerning husbandry, various sources, such as the papers by Bronson, et al. (23), Hoag and Les (47), Kallman (49), and Krzanowski (56) should be consulted.

C. PREPARATION OF MEDIA FOR OVUM CULTURE

Mouse eggs, being small, are often "lost" when large volumes of medium are used for their handling and culture. For this reason they are most often cultured in microdrops, which, however, have certain disadvantages. Among these are the rapidity with which they lose CO_2 and heat when removed from the incubator for observation, and the increase of tonicity through evaporation. In order to combat these perils, microdrop cultures are covered with a layer of paraffin oil of Saybolt viscosity 125/135, such as supplied by Fisher. Here another problem arises, in that many medium constituents are partly soluble in oil; even if their solubility is slight, composition of microdrop media may be altered drastically, due to the fact that the oil volume is so much larger than the medium volume. Consequently, the oil must be equilibrated one or two days beforehand with both medium and with CO_2. It will be evident that for studies in which the effects of different additives are to be tested, a separate oil batch should be equilibrated with each medium, complete with the additive in the concentration to be used. The only exception is for protein additives, such as bovine serum albumin. Because equilibration may lead to excessive foaming and to some denaturation, proteins are omitted from media used for oil equilibration. Good practice in medium preparation, therefore, is to add the protein last, after removing a portion to use for equilibrating oil.

An additional difficulty encountered in using oil layers is the experience that some lots are unsatisfactory, presumably because of trace impurities or because a given batch may induce spreading and flattening of microdrops. Hence, each lot should be pretested before use.

Equilibration is accomplished by bubbling a 5% CO_2-air mixture through a suspension of oil and medium. For this purpose, a cylinder of the compressed gas mixture is needed and a reducing valve. The oil may be autoclaved for 15 minutes at 15 lbs. pressure for sterilization, but a better method is to use dry heat, at 160° C. for one hour. In order to maintain sterility during gassing, the gas mixture may be passed through a membrane filter of the Swinney type (such as the sterile, disposable units supplied by the Millipore and Gelman companies) and led into the oil or medium bottle through an autoclaved Pasteur pipette attached to a short length of autoclavable plastic tubing, such as Tygon. The following procedure, illustrated in Figure 6, is satisfactory.

1. Add 30 ml. of sterile medium to a 500-ml. bottle containing 300 ml. of oil. Cap, and shake vigorously until a suspension is produced.

2. Assemble the gassing apparatus aseptically (see Fig. 6) and introduce the pipette so that the tip is in the aqueous layer at the bottom. This procedure should be carried out in a sterile hood if possible.

3. Cover the opening of the bottle with a flamed piece of aluminum foil and bubble vigorously for 15 minutes.

4. Remove the pipette, cap the bottle tightly, and shake vigorously for a minute or two.

5. Insert a fresh sterile pipette in the tubing and repeat the process.

6. After a final shaking, store in the refrigerator with the cap tightly closed. The suspension should separate overnight, and the oil is ready for use.

Oil should be equilibrated within a few days of use. Should there be doubt concerning the degree of equilibration, the bubbling process may be repeated at any time.

Modern media which support cleavage and early development of the mouse egg are derived from Whitten's original medium, based in turn upon Krebs-Ringer bicarbonate (108). Fortunately, mouse ova develop from the two-cell to blastocyst stage in defined media, which has made possible studies of the nutritional and other requirements of cleavage (8, 14-18, 21, 44-45, 65, 105-6, 109, 111). Since the media are defined, it is rather easy to vary con-

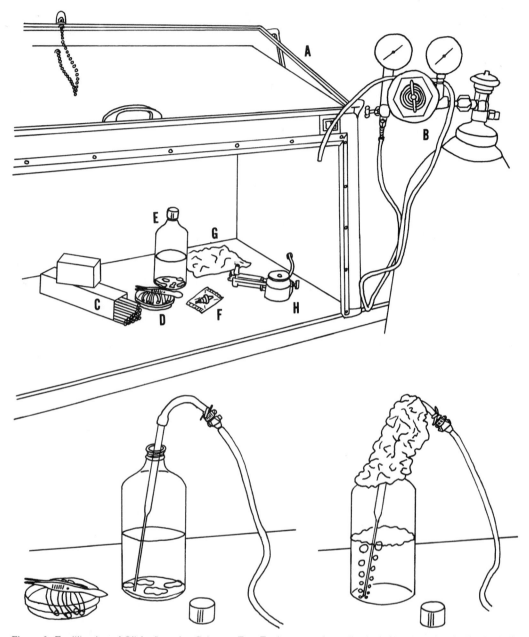

Figure 6. Equilibration of Oil for Layering Cultures. *Top:* Equipment and supplies include culture hood (A) with built-in ultraviolet lamp for sterilization before use; regulator valve on pressure cylinder containing 5% CO_2-air mixture (B); sterile Pasteur pipettes (C); short lengths of sterile tubing and forceps (D); bottle containing 300 ml. of oil and 30 ml. of nutrient medium (E); syringe-type membrane filter unit (F); aluminum foil (G); and burner for flaming (H). *Lower left:* Assembled equilibration apparatus. *Lower right:* Aperture covered by flame-sterilized foil. Gas mixture is bubbled through the medium layer on the bottom of the bottle.

stituents in order to study their effects upon early development, but it must be repeated that in order to conduct such studies, specifically equilibrated oil batches must be used.

The compositions of several media are given in the appendix. It will be noted that several anion-cation combinations are very poorly soluble in water, but their concentrations are low enough so that when media are properly made no precipitate forms. The trick is to dissolve salts which are likely to form precipitates in separate aliquots of water and then to add the solutions to the medium while it is being stirred rapidly. When all ingredients are added, the final volume is adjusted by addition of water.

A second important factor is the purity of ingredients. Reagent grade chemicals should be used at all times. When bovine serum albumin is included, a crystallized product, such as that sold by General Biochemicals, should be used. Water purity is equally important. Distilled water with resistivity of at least 7×10^5 Ohms, or water demineralized in a modern activated charcoal/mixed-bed ion exchange system, should be used. The critical nature of water purity also dictates precautions in washing and preparation of glassware, and one of the books dealing with tissue culture may be consulted for details. In general terms, however, such glassware should be put to soak in a non-toxic tissue culture detergent as soon as practicable after use and before it can dry out. After soaking, glassware should be brushed, rinsed well in tap water, and finally rinsed three times in the water used for preparing media. Glassware is stacked to drain dry, preferably on wire racks.

For further details concerning medium preparation, a text of tissue culture methodology, such as Parker's (see p. 85) should be consulted.

Part II

Laboratory Exercises

*1. The Preparation of
Capillary Pipettes*

Although the preparation of capillary pipettes is rapid and simple in principle, the student should practice the method until he is confident that he can expeditiously draw a pipette of the desired aperture. The pipettes are, of course, fragile, but with care one may last for hours of constant use without being broken or contaminated. When one is needed, however, it is usually needed immediately. Hence the ability to make pipettes rapidly and accurately means less delay in preparing material.

In learning to make capillary pipettes it is very worthwhile to have on hand a watchglass containing some eggs freshly flushed or fixed in formalin, since the usual objective is to make the aperture only slightly larger than the egg. Blastocysts, eggs in early cleavage, and eggs in cumulus, represent the range of sizes encountered and should be available. Ovarian oocytes may also be used, since they are easier to acquire (p. 21). Their size varies with the number of cumulus cells which remain attached.

If a burner with a small pilot flame is not available, a microburner should be made by softening a Pasteur pipette about an inch from the end and bending the segment 90°. The microburner so constructed is then attached to a gas outlet with tubing and the valve adjusted to give a flame about one-half inch long (Fig. 7). In constructing capillary pipettes, the thinned portion of a Pasteur pipette is held in the flame and rotated for a few seconds until the glass is quite soft. The pipette is then removed from the flame, and immediately pulled out

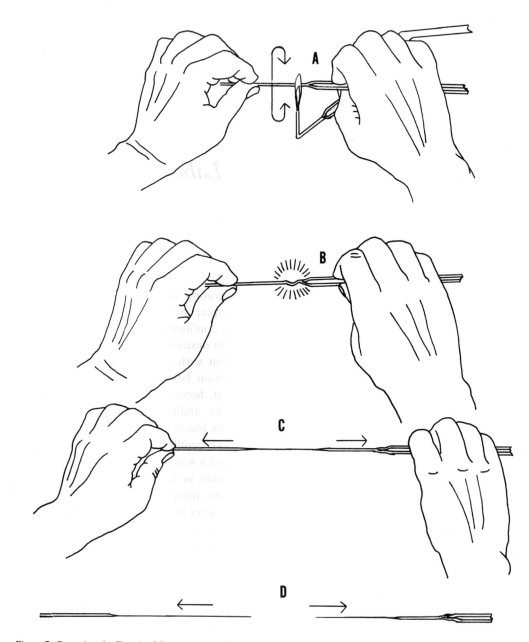

Figure 7. Procedure for Drawing Micropipettes. A Pasteur pipette is rotated in a small flame (A) until it is softened locally and quite easy to deform. It is pulled from the flame (B) and immediately drawn out a predetermined distance (C), depending upon the finished bore desired. The ends are held steady for a few seconds, until the capillary segment has cooled and the capillary broken by a sharp pull in opposite directions (D). The break thus formed is perpendicular.

into a capillary tube without breaking the capillary segment. The operator waits a few additional seconds for
the glass to harden and then he breaks the capillary segment with a snap by suddenly continuing to move the
hands apart. The capillary must not be broken by bending,
or a jagged aperture results; when the tube is broken by
longitudinal stress after hardening, the desired even perpendicular break usually occurs.

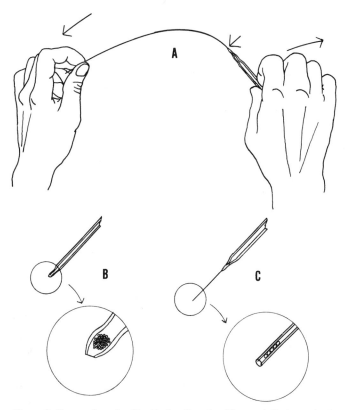

Figure 8. Preparation of a Pipette for Cumulus Masses. A Pasteur pipette
is drawn as for preparation of a capillary pipette (Figure 7) but not broken.
Instead, a file mark is made (short arrow in A) and the thin segment broken
off by pulling laterally. The result is a clean break of large aperture (B), compared with that of a capillary pipette (C).

Aperture diameter is controlled by the initial pull when
the glass is soft. For eggs without cumulus an aperture
of 80-100 μ is desirable, and this is achieved with an initial pull of about 10 inches. Eggs in cumulus require a
pipette pulled out about 4-6 inches. When handling cumulus masses containing 10 to 20 eggs it is usually nec-

essary to break the capillary by bending near the thick end, after making a file or diamond mark at the beginning of the expanded portion (Fig. 8).

For aseptic procedures, sterilized Pasteur pipettes are usually used, and capillary pipettes drawn as needed just before use. In practice, however, it is generally safe to use unsterilized pipettes taken directly from the box with a minimum of handling, provided the expanded portion is flamed just before the capillary segment is drawn.

Each student should have his own mouthpiece assembly and pipette adaptor, which may be made as shown in Figure 2.

2. Manipulation of Ova and Setting Microdrop Cultures

The operator should satisfy himself that he is adroit in manipulating ova before attempting to prepare cultures or perform experiments. Specifically, he must be able to transfer one or a predetermined number of eggs from one watch glass or culture dish to another in a volume of fluid only slightly larger than the egg itself (Fig. 9). When several eggs are to be transferred in the same pipette they must be drawn up close together and not widely spaced out, which means that the operator must maneuver them together and bring them into the pipette in close succession. For this purpose embryological watch glasses may be used, their curved bottoms simplifying the process of collecting the eggs together.

Since the pipettes are capillary in nature they are self-loading until the column of fluid rises to the expanded portion of the tube, and so to allow control of spacing, the pipette should be permitted to load itself to equilibrium before the eggs are picked up (Fig. 9).

Most culturing of mammalian ova is done in microdrops under oil (14). As an exercise, add oil to several disposable dishes to a depth of about 5 mm. and, using a capillary pipette, place three or four microdrops on the floor of each dish (Fig. 10). The drops should be approximately 3 mm. in diameter; this operation can be accomplished without magnification. Draw up one or more ova from the watch glass and, using the microscope, transfer them to the microdrops in the culture dish, taking care to introduce as little fluid as possible. When properly done, the volume of the microdrop is increased by only a small amount during addition of ova.

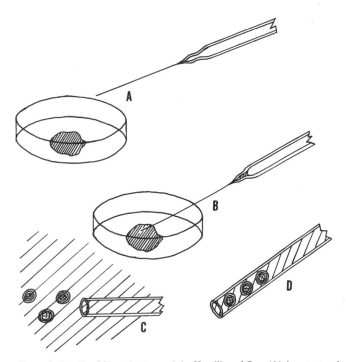

Figure 9. Loading Micropipettes and the Handling of Ova. (A) An empty micropipette is inserted in the dish of medium containing ova and allowed to
fill as far as possible by capillarity (B). Using the microscope, ova to be transferred are then grouped closely together (C) and drawn into the micropipette as close together and as near the tip as possible (D). The aperture of
a properly drawn micropipette is only slightly larger than the ova to be transferred. After the ova are loaded, the operator opens his lips to allow pressure
to equalize, and removes the tip of the pipette.

Some workers (e.g., Whitten, 108) prefer to avoid the
problems arising from microdrop culture by culturing
eggs in small plastic tubes. This method suffers in turn,
however, from the fact that it is difficult to see the ova
and to follow their development, and to transfer ova without loss, especially when small medium volumes are used.

B. THE MATURATION OF
OOCYTES *IN VITRO*

A remarkable observation is that oocytes taken directly
from the ovaries of some mammals can undergo apparently normal maturation *in vitro* (24, 28-29, 31, 50), although
attempts to fertilize such ova have not succeeded to date.
Recently, Donahue and Stern (29) found that mouse ova
maturate successfully in a high proportion of cases if a
defined medium containing pyruvate is used.

In the case of the mouse oocyte, the first maturation
division, with formation of the first polar body, takes
place 11 to 17 hours after setting cultures, and most oo-

cytes are found in evidently normal arrest at metaphase of the second meiotic division after 17 hours. Since arrested oocytes will remain in that condition for several additional hours, a convenient procedure is to set cultures at about 4:00 P.M. and to fix and stain them the following morning. Donahue's method (28) is described.

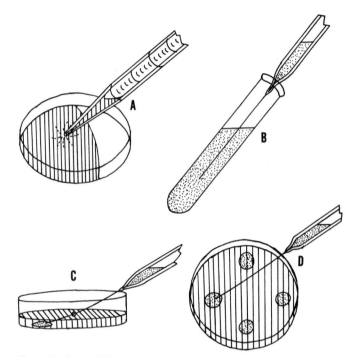

Figure 10. Setting Microdrop Cultures under Oil. Equilibrated oil (3-5 ml.) is added to a 60 mm. culture dish with a measuring pipette (A). A capillary pipette attached to a mouthpiece assembly is loaded with medium (B). In (C) the top of the pipette is pressed against the bottom of the dish under oil and a microdrop of 4-5 mm. is set. Three or four such microdrops are set in each dish (D), spaced equidistant from each other and away from the edge.

1. Preparation of Cultures

Paraffin oil must previously be equilibrated with oocyte culture medium (OCM, see appendix and instructions for oil equilibration in Part I), and this should be done at least 24 hours beforehand.

Mature or immature female mice are killed by cervical dislocation (Part I), since ether anesthesia, for example, interferes with recovery of oocytes. The ovaries are removed aseptically, trimmed of fat, and each transferred to an embryological watchglass. A small amount of fresh medium (approximately 0.5 ml.) is added to the watchglasses, followed by enough of the oil to form a complete layer. Using two needle holders fitted with 25-ga. needles, tear the ovary into four pieces. Then prick (do not tease) each piece in turn until the oocytes fall out into the watch-glass; some 30 oocytes should be recovered in this way; they are in cumulus, but this normally falls away with manipulation. The steps of this process are illustrated in Figure 11.

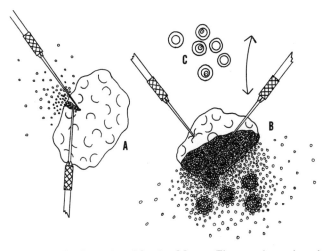

Figure 11. The Liberation of Ovarian OOcytes. The ovary is torn into about four pieces with needles (A). Each piece is then held with one needle and "pricked" with the other to release oocytes. The pricking motion (B) is vertical and avoids tearing or teasing. Large oocytes "fall out" into the medium. Only those bearing germinal vesicles (about 30 from each ovary) are used (C).

Cumulus cells adhere most strongly at the time of estrus, less so during diestrus.

Transfer the oocytes to another oil-layered watch glass containing medium, carrying as few cumulus cells along as possible. Examine each oocyte to determine whether a germinal vesicle is present; discard oocytes lacking them. During the manipulation any remaining cumulus and corona cells should fall away. It may be necessary to transfer the oocytes through one or two washes in oil-layered medium to get rid of all follicle cells. However, this procedure is only necessary when the effect of medium components on maturation is being studied, since follicle cells produce pyruvate and other substances (29).

After washing them, 6-8 oocytes may be transferred to each of several watch glasses containing about 0.5 ml. of equilibrated medium layered with (equilibrated) oil, and incubated at 37° C. in a gassed environment.

The culture dishes are conveniently set up beforehand and allowed to incubate until the oocytes are transferred to them, but they can be set up during the exercise if desired.

2. Staining and the Demonstration of Polar Bodies and Meiotic Metaphase

After 17 to 20 hours incubation, oocytes are removed from the microdrop culture system with drawn pipettes and transferred to a slide (about 5 per slide). All fluid is removed by sucking with the pipette; sticking of oocytes to the slide is improved if some drying is allowed to occur. Spots of wax-petroleum jelly mixture (20 parts Vaseline to 1 part paraffin of 56° C. melting point) are applied to the corners of a 22 mm.² coverslip, and this is inverted over the oocytes (Fig. 12). The coverslip is then pushed down slightly so that it holds the oocytes in place without deforming them. They are then fixed by introducing at one side of the coverslip a solution of glacial acetic acid (25%) and absolute ethanol (75%), drawn through for about one minute. The fixative is drawn off by touching the edge of the coverslip with a piece of paper tissue. Aceto-orcein stain is then substituted, and observations may begin after about a minute of staining. To bring the chromosomes into one plane, the oocytes may be crushed by applying pressure to the coverslip.

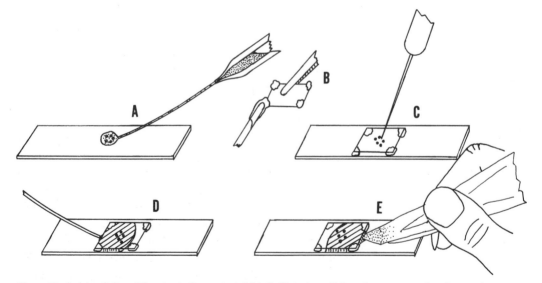

Figure 12. Staining Cultured Oocytes to Demonstrate Meiotic Metaphase. Cultured oocytes are placed on a microscope slide and most of the medium drawn off (A). A daub of paraffin-vaseline mixture is placed at the corners of a square coverslip (B) and the slip is inverted over the oocytes before drying occurs. The coverslip is pressed down (C) until the oocytes are just held but not deformed. Subsequently, fixatives, stains, and wash solutions may be added at one end (D) and drawn off at the other (E), using absorbent paper.

A stock solution of the stain is made up by dissolving 2.2 gm. of natural or synthetic orcein in 45 ml. of hot glacial acetic acid with continuous stirring. After 30 minutes, 55 ml. of 0.9% saline is added and stirring continued for a time. The stock solution should be made in a fume hood. A working solution is made up just before use by mixing two parts of the stock solution with three parts of saline and filtering at least once.

C. SUPEROVULATION AND PHASING OF OVULATION

Many years ago it was found that follicle-stimulating hormone, as contained in pregnant mare serum gonadotropin (PMS), and luteinizing hormone of human chorionic gonadotropin (HCG) could be administered in conjunction to accomplish two separate and important functions. First, they "phase" females by inducing estrus in about three-fourths of the cases some 56 hours after the beginning of injections; that is, when injections are begun on any of three days of the four-day estrous cycle they result in estrus and ovulation occurring on the third night after the first injection. Second, they increase the number of eggs ovulated from 12-15 to 20-40. Today this method is used almost universally in studies requiring the collection of mouse eggs, since it allows the collection of larger numbers of ova or the maintenance of smaller mouse populations. It was studied extensively by Fowler and Edwards (36). Superovulated eggs are regarded as normal ones since they produce viable young at the control incidence when transferred to foster females.

PMS and HCG preparations contain water-soluble hormones similar or identical to some anterior pituitary hormones. Thus, PMS mimics follicle-stimulating hormone (FSH) in stimulating follicle growth while HCG resembles luteinizing hormone (LH) in serving as the stimulus to ovulation. Both PMS and HCG are available as aqueous solutions. Two U.S. suppliers are Oragon, Inc. and Ayerst Laboratories. Working solutions are usually prepared by appropriate dilution to yield the desired dose in 0.25 ml. of distilled water or saline. The usual practice is to administer 5-10 I.U. of each preparation by intraperitoneal injection. The process is begun by injecting PMS at about noon on the first day, followed by HCG 48 hours later. Immediately thereafter, or late that afternoon, four to six injected females are added to each male's cage and plug-positive females are identified the next morning. Under ideal conditions the method gives successful results in 75% of cases as noted, but in practice 50% positive results are considered satisfactory. A final factor is apparent seasonal variation in dose response of females; it is good practice to give 5 units of each hormone in the fall and winter, and 10 of each in spring and summer.

Females may be "cycled," or induced to ovulate on a given day without superovulation, by giving doses of 0.5 - 1.0 units of each hormone.

D. RECOVERY OF OVA

The most effective method for recovery of ova depends upon the location of the ova, and this depends in turn upon the stage desired, since they move progressively down the oviduct and into the uterus as development proceeds. The time and stage relationship is given in Table 1, after the data of Lewis and Wright (57) and Whitten and Dagg (112). The table shows that there is a considerable overlap in developmental stages found at various locations and even at given times. Uncleaved zygotes are found in the ampulla up to one day after ovulation; cleavage stages up to eight-cell are located elsewhere in the oviduct up to about two days; and eight-cell and older stages are found in the uterus as well as in the tubes, or in the uterus exclusively beginning at three-and-a-half to four days. The data of the table are based upon naturally mated mice, whereas egg transport may differ somewhat in superovulated females, but the hormone injections do not seem to interfere seriously.

Table 1. Relationship between Time, Developmental Stage, and Location of Mouse Ova after Natural Matings

Copulation Age*	Ovulation Age (Est.)**	No. Cells	Location*
0 - 24	22 - 29	1	Ampulla
24 - 38	24 - 57	2	Upper oviduct
38 - 50	44 - 59	3 - 4	Mid-Lower Oviduct
50 - 64	50 - 59	5 - 8	Lower oviduct + uterus
60 - 80	77 - 80	Morula	Uterus
74 - 82	77 - 82	Blastocyst	Uterus

*Lewis and Wright (57).
**Whitten and Dagg (112).

The location and timing of desired stages is important, as will be seen, because different recovery methods must be used in some situations.

The basic method was first described in detail by Hammond (46). Females which have ovulated are killed by cervical dislocation, and the oviducts, along with a greater or lesser segment of uterus (see below), are removed. Before doing so, however, one should inspect the ovaries if ovulation was thought to have occurred a short time before (12 hours or less), since bloody fluid within the ovarian capsule or small clots upon the ovaries are often readily visible with the unaided eye and confirm recent ovulation. At this stage, inspection of the ovary with the dissecting microscope should also reveal ruptured Graafian follicles. At somewhat later stages inspection reveals little, since the follicles heal rapidly, and yet several days are required for gross recognition of corpora lutea.

In removing oviducts one should first examine the genital tract at low magnification and learn to recognize these small structures with the unaided eye (Fig. 13). With a little practice it becomes possible to remove the oviduct alone in a few seconds, using only scissors and forceps and without magnification (see below).

If the uterus and oviduct are to be removed together, the former is first severed near the cervix on one side, and the cut end is raised somewhat and put on a stretch with the forceps. The mesentery of the uterus is then cut away close to the uterine attachment so that all fat is removed. The cut is carried forward past the oviduct, making sure that the coils are not cut but that all fat is trimmed away; one of the chief values of a preliminary inspection with magnification is to enable one to learn to distinguish fat, ovary, and coiled oviduct, especially since fat carried into the collecting dish interferes with recognition and recovery of ova after they are liberated.

Having removed most of the mesentery and all of the fat, the stretched tract now consists of ovary, oviduct, and uterus in rostral-caudal series (Fig. 14). The uterus is now transected, but the point of transection is determined by the location of the ova within the tract. If the females ovulated less than two-and-a-half days beforehand, all ova are usually found in the oviducts. Hence, the transection in this case is made about 1 mm. below the utero-tubal junction, since the small piece of uterus is convenient for orientation purposes without being bulky. If ovulation occurred two-and-a-half days or more beforehand, many or all ova will be found in the uterus,

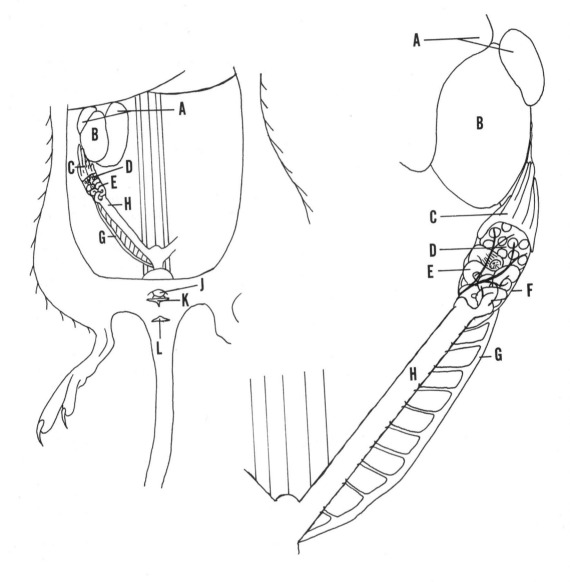

Figure 13. Female Internal Reproductive Organs. A. Perirenal fat; B. Kidney; C. Periovarial fat body, with peritoneal attachment at upper end; D. Ovary; E. Oviduct; F. Ovarian bursa; G. Uterine blood vessels; H. Uterus; J. Clitoris; K. Vagina; L. Anus.

which must also be flushed. In this case, the uterus is transected near the cervix.

Following transection, the rostral portion of the uterus is again stretched with the forceps, and the V of a small scissors worked slowly between the oviduct and the ovary. By maintaining the stretch and working the scissors slowly, the mesentery between oviduct and ovary is gradually cut and the two organs slowly separated until the membrane may be safely cut clean through. Magnification is not required, and cutting into the oviduct seldom occurs.

The severed oviduct and attached uterine piece are transferred immediately to medium or saline to prevent drying.

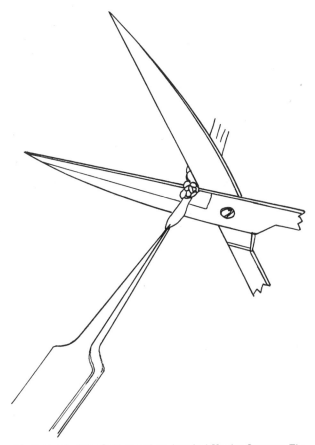

Figure 14. Removal of the Oviduct and an Attached Uterine Segment. The mesenteric attachment of uterus, oviduct, and ovary is cut, excess fat trimmed away from the tubal portion, and the uterus transected. The cut end of the uterus is grasped by forceps and the three organs stretched and kept under tension, the ovary remaining attached. Scissors are inserted between oviduct and ovary as shown and "worked" slightly while the tension is maintained, until uterine segment and attached oviduct are freed. A microscope is not necessary.

2. Liberation of Ovulated Ova and Zygotes in Cumulus

Within about 12 to 14 hours after ovulation (that is, the morning after mating), ova are found within the ampullary portion of the oviduct, embedded in cumulus and usually stuck together as a single large mass of ova and cumulus cells. An effective way to remove such cumulus masses from the ampulla is to simply tear the expanded ampulla open and allow the mass to float out into the medium. Using the dissecting microscope, locate the ampulla, and, taking advantage of its near-transparency confirm the presence of a cumulus mass and note that ova are included within it. The ampulla may be located on one or the other side of the coil, or at a margin; rarely is it covered by lower coils of the oviduct. It is identified by its expanded character at this stage and by the prominence of the longitudinal epithelial folds. The beginner should also take the opportunity to locate the tubal ostium which is often hidden by coils (Fig. 15). It is immediately adjacent to the ampullary expansion.

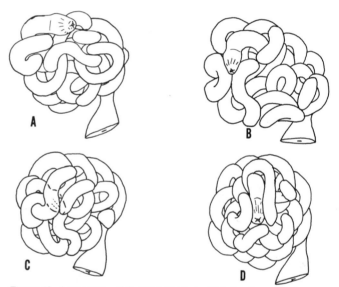

Figure 15. Appearance of Isolated Oviducts. Tubal ostium may be superficial, as in A and B, and relatively easy to locate. When completely or almost completely hidden (C and D) it must be searched for in the coils. In this case especially, a remnant of bursal membrane may be left reflected from the ostium and may sometimes cover it. Since it is transparent, it is detected by impedance of flushing needle entry. Note that there is a characteristic sharp bend between ostium and ampulla. During estrus the ampulla is greatly expanded (not shown).

For liberation of the intact cumulus mass use a pair of watchmaker forceps and a disposable 25-ga. needle attached to a needle holder or to a 1 ml. syringe as a handle. Seize the caudal portion of the expansion with the forceps and make a tear between forceps and cumulus mass; making the cut below the mass takes advantage of the direction of ciliary activity in liberating it. The tear may be pulled further open with two pairs of forceps, and in most instances the mass is readily extruded. If not, the tear may be enlarged still further or the mass expressed by gently compressing or stroking the upper portion of the ampulla.

Before transferring the mass to another dish or to another drop in the same dish, however, a caution should be noted. The cumulus mass is too large to fit without distortion in the usual capillary pipette. Further, the cumulus is quite sticky, and likely to remain stuck within the pipette. In order to avoid this, a pipette should be made so that as little contact as possible occurs between cumulus and glass. This is accomplished by constructing a capillary pipette in the usual way, except that the break is made by bending the cooled capillary after making a file mark (Fig. 8, p. 17). The result is a smooth and rather wide aperture immediately below the flared portion of the pipette. Such a pipette should be well filled with medium before the mass is picked up to avoid the unusually tenacious adherence that occurs when cumulus cells (or any other) contact dry glass. With a little manipulation the cumulus cells normally fall away and are lost. The cells may also be removed by hyaluronidase (p. 32).

3. Flushing Ova from Oviducts and Uteri

Although this is one of the more difficult procedures to be described, it can be mastered within an hour or so with proper instruction and careful attention to detail.

A first consideration is the alteration of syringe needles to render them suitable for flushing. Needles of 30-ga. aperture are used. With the aid of a dissecting microscope and an abrasive stone, they are blunted. If necessary (and it usually is), the shaft of each needle must be ground off to reduce its diameter enough to allow the needle to enter the ostium of the oviduct. The procedure is illustrated in Figure 16.

Plugging is a considerable problem with fine gauge needles. After preparation, needles should be washed by squirting detergent solution through them and by running a cleaning wire through the aperture. They are then placed in protective glass needle holders which are stop-

pered and autoclaved. Needles can often be cleared by
holding them in a flame.

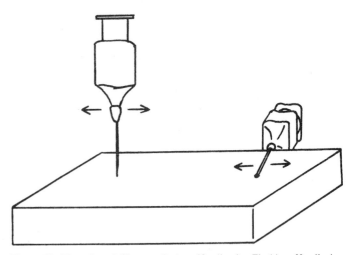

Figure 16. Alteration of 30-gauge Syringe Needles for Flushing. Needle is
blunted on an oilstone *(left),* and, if necessary, the external diameter is reduced by honing with lateral movement while rotating the shaft *(right).*

As Table 1 illustrates, stages of two-and-a-half days
or greater ovulation age are often found in the uterus.
In this case, both oviduct and uterus are flushed, and so
the initial cut in removing the organs should be made
close to the cervix and the two removed together. They
are transferred to a drop of medium (as small as practicable so that it will be unnecessary later to search through
a large volume of fluid for the flushed eggs), and the two
sides of the tubal coils examined in turn in order to locate the ostium. This structure is most often located near
the center of the coils, either apparent near the surface
or, frequently, hidden by coils (Fig. 15). In the former
case, it is usually easy to identify once one is familiar
with its appearance. If hidden, however, it must be looked
for. Possibly the best procedure for doing this is to locate
the ampulla first and follow it to its terminus at the ostium.
Most often, the ampulla is marginally located, but at later
postovulatory stages it is less expanded and therefore
less conspicuous. Using needles or forceps, suspected
segments should be partly freed and the terminus sought
for. In most instances the ovarian portion of the ampulla
makes a rather sharp bend in accord with the normal
central location of the ostium (Fig. 15). When the ostium
is located it may be entered for a very short distance with

the fine needle to be sure that a fold of the transparent ovarian capsule does not lie over it.

A flushing needle is attached to a small syringe which is loaded with 0.1-0.2 ml. of medium. If the operator is right-handed, the syringe is held in that hand, and watchmaker forceps or fine needle is held in the left. The ampulla is seized or pinioned against the bottom of the dish

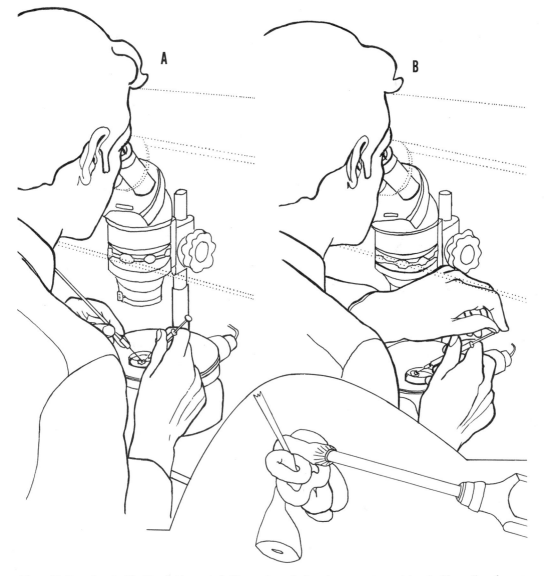

Figure 17. Procedure for Flushing Oviducts. A. Oviduct and attached uterine segment are pinioned with needle or forceps held in the left hand, with the ostium pointing toward the right *(inset)*. The flushing needle, mounted on a syringe held in the right hand, is inserted and the oviduct pinioned against the bottom of the dish by the needle end from within (A). The holding needle is then put down, and the left hand is crossed over to operate the syringe (B). A small amount of fluid (about 0.05 ml.) is flushed through and the resulting drop examined for ova.

so that the ostium points toward the operator's right and is visible. With the operator's hands steadied on the hand-rests, the flushing needle is inserted into the ostium for a short distance, or as far as it can be practicably placed. The tip of the flushing needle is now pressed gently against the bottom of the dish with enough pressure to hold the oviduct in place, but with little enough to avoid tearing the tubal wall. The left hand now puts down the forceps or fine needle and crosses over to the right in front of the operator, while the flushing syringe and needle are held steady. With the fingers of the left hand the syringe piston is depressed (see Fig. 17 for steps). The operator should confirm expansion of the tube as fluid enters it, followed by the flow of fluid out of the uterine end. The tube may now be discarded and the fluid searched for ova.

E. REMOVAL OF CUMULUS WITH HYALURONIDASE

In tests of nutritional factors required for cleavage it is essential to remove cumulus cells, since they have been shown to provide pyruvate or other compounds which support cleavage. A solution containing 300 units/cc. is prepared by dissolving 20 mg. bovine testis hyaluronidase and 200 mg. of polyvinylpyrrolidone (PVP) in 20 cc. of Dulbecco's phosphate-buffered saline (PBS) (7). The solution is filter-sterilized (Millipore HA) and distributed in 1.0 cc. ampules for freezing. A large drop of the enzyme solution is placed in a petri dish, and eggs in cumulus are transferred to it. After a few minutes the eggs may be pipetted in and out a few times to help dislodge the cumulus cells. Incubation at 37° C. may be necessary, but in most instances the cells detach readily after a few minutes at room temperature. Medium droplets may be placed at other locations in the dish and the denuded ova passed serially through them to wash and separate them from cumulus cells.

F. OVUM CULTURE

In planning exercises or experiments with the culture of ova, one should bear in mind the special nutritional requirements of different cleavage stages as well as the two-cell block phenomenon. Although no extended discussion of these factors will be attempted, their mention is necessary. Before planning such exercises the operator should familiarize himself in some detail with the interesting nutritional features of mouse cleavage. Some helpful reference sources are those of Biggers, et al. (8), Brinster

(16-18), Gwatkin (44-45), and Wales and Whittingham (106). Mouse ova were first cultured in defined media by Whitten (108); McLaren and Biggers (61) demonstrated the normality of such ova by rearing the young produced following foster transfer.

The two-cell block phenomenon refers to the fact that eggs which have not completed the first cleavage at the time of flushing do not generally develop beyond the two-cell stage when they are cultured, unless oviduct fragments are present (115-16). However, eggs which are removed after the first cleavage is completed can be cultured with quite good success until the normal implantation stage (i.e., late blastocyst), provided that required metabolites are present in the medium. Experimentally, the two-cell block has recently been overcome by Whitten and Biggers (111) using certain genetically defined mouse strains and slightly hypotonic medium. Because of possible difficulties in obtaining the required inbred mouse strains, their procedure is not described in detail since it is probably unsuitable at present as a standard laboratory exercise.

As noted, ova recovered in the two-cell or later stages continue to cleave normally, as shown by the birth of normal young following transfer of cultured blastocysts to pseudopregnant foster females. However, some of the enzymes of the oxidative metabolism of two-carbon fragments appear to be lacking or inoperative early in cleavage, making their appearance sequentially. Thus, pyruvate is required for further cleavage of the two-cell egg, while the eight-cell egg develops well using glucose alone. Unless nutritional factors are being studied, therefore, a medium containing both glucose and pyruvate will support cleavage of all two-cell and later stages. Composition of a medium used by Biggers, et al. (7), is given in the appendix. Lactate is also present in this medium. Although not necessary, it can substitute for pyruvate at most stages of development, and the medium seems to be empirically more effective when it is present in addition to pyruvate.

Mulnard (76-77) has also had considerable success in culturing mouse ova, using a medium based on Krebs-Ringer saline, as did Whitten. Although Mulnard's medium lacks added two-carbon sugars it is supplemented with mouse serum, which is said to be essential for cleavage.

In Mulnard's system, eggs are cultured in microtubes of 1.5 mm. internal diameter, nine-tenths filled with medium and immersed in paraffin oil in such a way that a drop of air is present at each end. The size of the air drop is critical (tubes three-fourths filled or completely filled with medium do not permit development), suggesting that too much as well as too little oxygen is harmful. The small medium volume used may also act to permit development in the presence of comparatively small amounts of one-carbon sugars through lessening of diffusion effects.

Preparation of Mulnard's medium is given in the appendix.

A consideration of crucial importance is the removal of all cumulus cells whenever ova are to be cultured with an eye to the study of nutritional components essential to cleavage and normal development, since Donahue and Stern (29) have shown that these cells produce pyruvate and can in fact substitute for pyruvate in media lacking it. In all such experiments, therefore, it is essential to remove all follicular and cumulus cells by pipetting, followed by transfer through several microdrops of medium for washing purposes. Before ova are dispensed to cultures, the microdrops containing them should be inspected carefully at high magnification to be sure that all adventitious cells have been removed.

1. Microdrop Cultures

This procedure is described in section A, page 18. The operator should remember to set microdrops 1-2 cm. apart, since some spreading of drops occurs after 12 to 24 hours. The amount of drop-spreading is a function of the nature of the culture dish surface and of the particular medium being used. Overspreading of microdrops should be avoided, partly because of the risk of adjacent drops merging, and partly because undue spreading can lead to thinning of the medium layer in the drop to the point where the ova are compressed by the oil layer or caught up in it altogether.

Equilibration of the layering oil is quite essential whenever nutritional factors are being studied, for reasons given in an earlier section. Equilibration should be completed at least one day before the cultures are set. In addition, it is very good practice to set up the microdrop culture dishes with oil and medium the day before ova are introduced, incubating them in the interval in a 5% CO_2 atmosphere. In this case, of course, the oil will have been equilibrated two days beforehand.

It is customary to store CO_2-equilibrated medium and oil at 5° C. In any case, oil or medium should be regassed in advance of use whenever there is any suspicion that CO_2 has been lost. As noted earlier, dissolved CO_2 is in equilibrium with bicarbonate, which appears to be a desirable component of the medium. In addition to causing depletion of bicarbonate, the evolution of CO_2 results in a drastic rise in pH which in turn is harmful both directly and by virtue of increased tendency toward inactivation of medium components under these conditions. Therefore, if the pH can be monitored continuously, the risk is greatly lessened. The most convenient way to accomplish this is through the addition of phenol red to the medium, to 0.005% (sterile solutions of 0.5% are available from suppliers of tissue culture media). Since the dye is essentially insoluble in oil, medium containing it may be used for oil equilibration. It does not appear to be toxic to cleaving mouse ova or to yield toxic products of photoreaction when cultures are incubated in the light.

A final caution concerning oil equilibration bears repeating: media used for this purpose must be free of serum albumin or other protein, since excessive foaming and possible denaturation may result. Hence each different medium used must be made up with and without BSA, and a different batch of oil equilibrated with each kind of the protein-free media.

During manipulation and the setting of cultures it is very difficult to prevent a temperature drop at least part way to ambient. Fortunately, temperatures of about 20° C. or higher do not seem to be detrimental to the ova, provided they are not prolonged for more than an hour or so. If longer delays are required, ova should be incubated in the interim. Embryological watch glasses, which have relatively massive bases to act as heat sinks, may be preincubated and used for temporary holding. As a precaution against drying and CO_2-loss, such holding dishes should be layered with oil. In any event, experiments should be planned so that flushed ova are set in culture and incubated as soon as possible.

2. Tube Cultures

The simplest way of setting ovum cultures is to place them in previously prepared tubes containing medium (0.5 ml. or less). Tube cultures are usually not oil-layered, since they can be removed one at a time from the incubator, seeded, and quickly returned. Sterile plastic tubes of about 12 × 75 mm. (such as Falcon #2003) are suitable.

Although this method is quite effective, it suffers from the considerable disadvantage that monitoring the cleavage process is very difficult because of optical problems. An inverted microscope is required for observation within the tubes, and, since resolution is poor when viewing is done through the curved wall, it is quite difficult to distinguish normally cleaved from fragmented or otherwise abnormal ova. Another result of the use of comparatively large volumes of medium, combined with difficulties in observation, is that ova tend to be "lost," and recovery in high proportion is much more difficult.

The tube method is very convenient, however, when large numbers of eggs are to be cultured simultaneously, as, for example, in most experiments which require extraction of specific components. Unless it is necessary to culture hundreds of eggs, therefore, the microdrop method is preferable.

3. The Recognition of Abnormal Cleavage

In spite of careful husbandry and technique, one may sometimes observe large numbers of cultured ova which are of dubious normality, probably due often to damage occurring *in vivo*. In any experiment, however, it is important to assess the proportion of abnormal ova as well as may be. For this purpose even transfer to foster females, inordinately laborious and expensive as a routine process, is impractical, since even in the best circumstances only a proportion complete development after foster transfer. Hence one is forced to use simple morphological criteria, inadequate as they are, to assess normality.

The most frequently observed abnormality of cleavage, and the one most difficult to assess, is fragmentation (Fig. 18)—a chaotic, disorganized cleavage. One criterion of fragmentation is the occurrence of blastomeres of grossly unequal size, but this calls for careful judgment since there is some normal disparity in blastomere size, especially at the three- and four-cell stages and in the morula. Cells of the blastocyst inner mass are normally larger than those of the trophoblast but are not readily

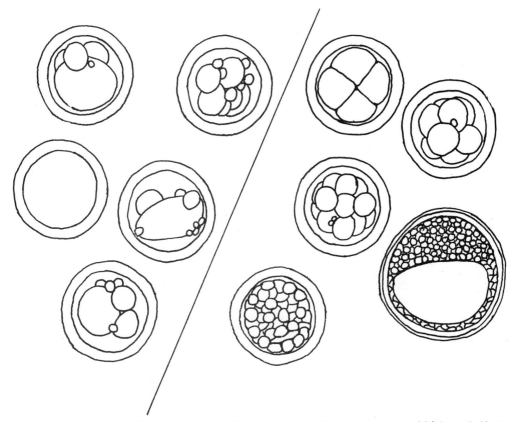

Figure 18. The Appearance of Normal and Abnormal Ova after Culturing. Normally cleaved ova *(right)* contain blastomeres of roughly equal size, forming an organized whole. Abnormal cleavage results in fragmentation, in which blastomeres vary greatly in size, often are not round, and sometimes lack chromatin. Empty zona membranes, presumably the result of complete cytolysis, are occasionally seen.

seen in the dissecting microscope and in any case are recognized as normal when confined to the inner cell mass.

A second criterion of fragmentation is the appearance of more blastomeres than appropriate for the time course of the experiment.

A third criterion is the absence of chromatin from some blastomeres, but this is less often applied since staining is required.

A second kind of cleavage abnormality concerns the arrangement of otherwise normal-appearing blastomeres in unusual configurations. Early in cleavage, blastomeres are arranged in well-defined tiers, giving each stage a characteristic appearance (Fig. 18). It is important to know the normal appearance and to recognize bizarre arrangements.

Blastocysts often present an abnormal appearance,

most frequently as a vesiculated inner cell mass or double blastocyst cavity. Unusually small cavities in relation to time *in vitro* are also frequently seen, but probably are often normal. The blastocyst has been shown to pulsate during development. Further, blastocyst expansion is quite rapid when it begins, and so one may reasonably expect development of the cavity to vary normally within rather wide limits.

It is important in assessing cleavage abnormalities to do so as soon as possible after removal from the incubator. In particular, the appearance of blastocysts is somewhat labile, degenerative changes (especially vesiculation) being recognizable after an hour or so at room temperature. Although foster transfer experiments suggest that many such ova may not be irreversibly damaged, their occurrence makes for much difficulty in assessing the success of experiments, and so rapid scoring is essential.

Occasionally, empty zona coats are seen, often in large numbers and sometimes under circumstances which suggest complete autolysis of ova within the zona. After about the fourth day postovulation, cultured ova "hatch" from the zona in culture, leaving membranes which are split open, plus rather nondescript cell masses which degenerate quickly. Although it is not certain whether hatching occurs *in vivo* (opposed to gradual disappearance of the zona before implantation), it is a regular happening in culture and does not denote abnormality.

It should now be apparent that comparison of results from experiments can be quite misleading unless statistical methods are used. The application of such methods to this problem is discussed at some length by Biggers and Brinster (4), and by Biggers et al. (7).

G. THE MICROINJECTION OF EGGS

Lin (58) has injected mouse zygotes with a saline-protein solution and obtained morphologically normal development at 17 to 19 days gestational age after foster transfer in 11% of cases. Because the method is complex and requires specialized equipment, it will not be described in detail here. However, the method is a promising tool, and a brief outline is desirable.

Hybrid zygotes were obtained by flushing oviducts and transferred to medium in a vaseline well on a microscope slide under oil. A holding pipette of 15-μ bore, used with a Leitz micromanipulator, was attached to a screw-operated syringe. With it, zygotes were held firmly against the end of the pipette by negative pressure. Injections were made with an especially constructed capillary pipette of 1 μ diameter; the injected volume was determined by calculation from the distance the meniscus moved within the capillary, using linear measurement from a superimposed ocular micrometer. Measurement and control of the injected volume were sufficiently precise that eggs could be injected with volumes small enough that noticeable swelling did not occur.

Transfers were made to the ovarian capsule of pseudopregnant females (p. 48). Although survivors were essentially normal near the end of gestation, one-fourth were somewhat smaller than controls.

H. THE TRANSFER OF OVA TO FOSTER FEMALES

Although a meticulous operation, ovum transfer has been well described and has been performed in many laboratories (6, 26, 35, 38-39, 52-53, 61-63, 72, 96, 113-14). An unexplained feature of the process is a rather constant but moderately low success rate, even in the hands of experienced workers. Thus, many laboratories experience success at about 50% by the criterion of term fetuses. Some laboratories, however, have had somewhat better results (72), approaching values which are probably about as high as the undisturbed mouse itself produces.

Successful transfer requires rather precise phasing with respect to postovulation age of the donor ova and pseudopregnant age of the recipient in conjunction with route of transfer. Thus one- and two-cell ova (but not later stages) can be transferred beneath the ovarian capsule of half-day pseudopregnant females (88-90), and three-and-a-half day blastocysts (but not earlier stages) can be transferred directly into the uterus of two-and-a-half to three-and-a-half day pseudopregnant recipients, or any cleavage stage may be transferred into the ampulla of half-day recipients (38-39). Thus, careful advance planning is required.

Pregnancy or pseudopregnancy must be established in recipient females at or before the time of transfer, depending upon the stage of ova to be transferred, as noted above. Recipients are mated with normal males if it is desirable to produce both donor and native eggs in the same female. In most experiments, however, the development of native eggs tends to interfere, and so vasectomized males are used to establish pseudopregnancy. In either case the investigator should use different genetic combinations, so that donor and native embryos may later be distinguished phenotypically. If Swiss albino donors and recipients are used, for example, C57Bl males may be used for recipient matings. Any native eggs that develop in the recipient are thus detectable after 17 or more days postcoital age, on the basis of iris pigmentation.

Uniparous females should be used as foster recipients.

1. Vasectomy of Males for Preparation of Pseudopregnant Foster Females

Males to be used for pseudopregnant matings should be 12 weeks of age or older, up to about one year of age.

Mice may be anesthetized with ether or nembutal, but in either case an ether cone should be at hand for "touching up" (p. 10). Mice that are anesthetized deeply enough not to require some additional anesthesia during the operation are often overanesthetized and die. Unless the operator is quite experienced it is far better to underanesthetize and supplement anesthesia than to lose the time spent in the operation.

A mouse is restrained in the supine position and the scrotum wiped with 70% alcohol. Using forceps and scissors, a longitudinal incision of about one-half inch is made in the scrotal midline (Fig. 19). The ductus deferens is now dissected on each side in turn and severed. Avoiding sharp instruments, the operator bloodlessly dissects one scrotal sac toward the inguinal canal, since the testes are almost invariably retracted within the canal, which appears plugged with fatty tissue. This is seized and pulled down, bringing the testis with it, followed by the spermatic cord, which is as large as the testis itself. Maintaining the stretch, the fatty cord is dissected with forceps and the ductus deferens searched for; it is identified by the fact that it is much larger than any other structure in the cord. Until the operator is familiar with its appearance he should perform rather complete dissections of the cord to avoid misidentification and subsequent cutting of a nerve or vessel. Once the ductus is seen, there is little doubt of its identity, and subsequent operations may be performed swiftly.

Following location of the ductus, dissect free as much of it as possible, maintaining the stretch on the lower pole of the testis. After dissecting free at least half an inch of the ductus, cut it with scissors as close as possible to the inguinal canal, without ligation. Identify the severed distal portion and cut this at the tail of the epididymis, removing the half-inch (or longer) segment. The scrotal contents are then repositioned within the sac and the operation repeated on the contralateral side, working through the single midline incision. When the operation is completed a single wound clip is used to close the incision. It is usually not necessary to remove the wound clip later.

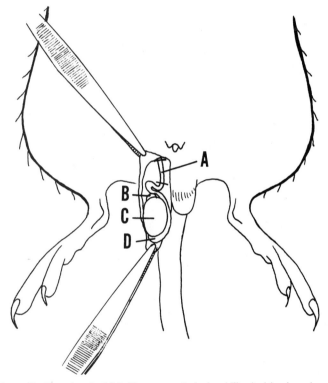

Figure 19. Procedure for Male Vasectomy. A single midline incision is made, contents of one scrotal sac pulled out, and the spermatic cord dissected. The vas deferens is identified and followed to the epididymis. It is cut in two places (without ligation) so as to remove the longest possible segment. The operation is repeated on the other side through the same incision. A. Vas deferens; B. Head of epididymis; C. Testis; D. Tail of epididymis.

Although mice are moderately resistant to infection one should always use instruments that are at least clean and preferably sterilized. The alcohol wipe-down before and after the operation is helpful, but the use of antibiotics is unnecessary and probably not very effective unless administered repeatedly for several days. In practice, serious infection seldom results from this operation or from laparotomy.

Vasectomized males should be allowed a week to recover (without access to females) before being used for pseudopregnant mating. Since the operated animals are more valuable than ordinary males, it is good practice to select proven breeders for the operation, but quite satisfactory results are normally obtained in any case.

It should be remembered that vasectomy does not interfere with mating or with libido, since the testis is still present and functional; sperm continue to be produced but are resorbed in the epididymis without consequence to other functions. In addition, the seminal vesicles, coagulating glands, and other secretory glands of the male genital tract are distal to the excised segment of the ductus, and so vaginal plug formation occurs normally.

Mating with recipient females is accomplished in the usual way. Recipients are often injected beforehand with two units each of PMS and HCG in order to phase estrus and increase the number of plug-positive recipients. However, Mintz (72) feels that hormone treatment of recipients slightly reduce the proportion of successful transfers.

2. Transfer of Blastocysts Into Uterine Horns

The method described here is based upon the work of McLaren and Michie (62-63), and its utility has been extensively demonstrated. A detailed description of a similar method is given by Mintz (72): Blastocysts of three-and-a-half days postovulation age, taken either from culture or flushed directly from pregnant females, may be transferred into the uterine lumen of two-and-a-half day or three-and-a-half day pseudopregnant recipients. McLaren and Michie use albinos as donors and C3H $\times$ C57B1 F_1 female hybrids as recipients. Thus, a genetic marker and hybrid vigor, which seem to increase the number of "takes," are combined. The incidence of success is further increased by using as recipients "proven" or uniparous females—that is, recipients which have previously borne a litter. All of these conditions should be fulfilled if success on the order of 50% is expected; experience is also an important factor.

For this procedure, two operators are almost essential. One is designated to prepare the recipient, and the other to collect blastocysts and perform the transfers.

a. Preparation of the recipient

In the majority of experiments it is desirable to inject only one uterine horn as a further control for the detection of native implantations; crossing over from one horn to the other occurs rarely. In either case, both horns may be reached via a single midline incision of the mid-dorsal skin, beginning at the level of the lowest rib and extending caudally for 3 cm. or so. The opening may then be retracted to either side to reach each horn in turn. The fat pad covering the ovary is usually visible through the thin lateral abdominal wall which is incised for about 1 cm. The fat pad is seized and exteriorized through a sterile slitted gauze pad, bringing the oviduct and uterus with it. The operator must now orient the recipient and its uterine

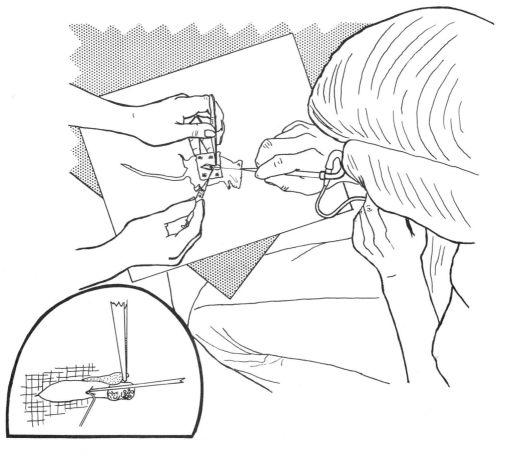

Figure 20. Transfer of Blastocysts into Uterine Horns. The uterus is exteriorized through a gauze square and stretched. It is punctured with a bent syringe needle mounted on an empty syringe, and the needle tip introduced into the lumen. The needle is withdrawn almost entirely, and the puncture hole is held open by retracting the needle laterally. The second operator then inserts the egg-bearing micropipette through the opening thus formed. Stippled area in inset is the fat body, convenient for grasping when maintaining tension on the uterus.

horn so that insertion of the blastocysts is convenient
for the second operator. In order to accomplish this, an
unusually narrow table (not more than three feet wide)
should be used, or the corner of a table. A typical arrangement is indicated in Figure 20.

When his partner is nearly ready for the transfer, the
recipient operator prepares the uterine horn. With forceps
held in one hand and grasping the horn near its cranial
end, he puts the horn on a stretch. With the other hand,
using a 30-ga. syringe needle, he makes a diagonal puncture in the horn, beginning about half a centimeter from
the cranial end and extending caudally. In making the
puncture, the recipient operator must be sure of entering
the uterine lumen without lacerating the epithelium.
To this end, the needle should be inserted in such a way
that it can move freely within the lumen, but the operator should avoid "testing" the freedom of movement ex-
cessively, since this process encourages laceration of the
endometrium. The hallmark of the beginner is to pene-
trate too shallowly, so that the injected blastocysts lodge
within the endometrium. Jabbing is to be avoided at all
costs. Entry should be made smoothly and firmly, with
the operator watching the angle of the needle and marking
the depth of penetration so that an estimate of the location of the tip is possible at all times. As it becomes evident that the lumen is being approached by the needle
tip, the course of penetration should be altered so as to
be more parallel to the horn.

When the operator feels, from the position and depth
of the needle, that the lumen had been entered, then—
maintaining a good stretch on the horn—he should withdraw the needle slightly. If it slides easily, the lumen has
been entered successfully; if not, the tip is probably within
the endometrium or myometrium.

The recipient operator and his partner must be in constant communication concerning their progress. The
lumen should not be penetrated until the donor operator is nearly ready. When the penetration is made, the
recipient operator, maintaining the stretch on the uterine
horn, withdraws the needle almost, but not quite, completely (Fig. 20); he then pulls the needle tip laterally
so as to make an aperture which is visible to the donor
operator.

Now the convenience of the donor operator must

be considered. The recipient operator must point the stretched uterine horn toward the micropipette held by the donor operator and hold his hands to the side to prevent obscuring the view of the donor operator.

b. Preparation of donor blastocysts

When direct transfer of blastocysts from donor to recipient is required, the donor operator should begin collections somewhat in advance, since a longer time is needed to flush oviducts than to prepare recipients. Whether blastocysts from intact animals or from culture are to be transferred, additional time must also be allowed for scoring and for selecting normal-appearing blastocysts. Fortunately, removal to room temperature for an hour or so does not seem to impair the ability of blastocysts to develop normally after transfer.

Three or four blastocysts are usually transferred to one horn. These are collected in the micropipette, all close together near the tip, as soon as the recipient operator indicates that he has opened the lumen of the horn. An important factor at this point is to be sure of loading the pipette with the capillary volume first so that fluid will not continue to enter it after the blastocysts are picked up (Fig. 9). With three to four blastocysts loaded near the tip, the donor operator removes the pipette from the culture microdrop and, with as little motion as possible, aims it in the direction of the prepared recipient horn. While his partner holds the puncture hole open with the tip of the syringe needle, he gently urges the tip of the pipette into it, allowing it to pass two or three millimeters within the lumen.

The blastocysts are extruded in as small a volume as possible. A convenient way of accomplishing this is to watch the top of the fluid column in the pipette while increasing the air pressure in the pipette; as soon as the top of the column is seen to move at all, the operator should stop expelling and withdraw the pipette. In order to determine whether all of the blastocysts have been extruded, the remainder of the fluid should be forced out under oil and the droplet examined for unexpelled blastocysts. In transfer operations this is an important part of the experimental record.

Upon completion of the transfer, the horn, oviduct, and ovary are replaced carefully within the abdominal cavity and the skin incision is closed with two or three wound clips. The muscular incision is not sutured or clamped.

From the experiences of a number of laboratories this method may be considered very effective. Some experience is required, however. Barren recipients are usually those that suffered "traumatic" operations with excessive bleeding, pinching of viscera, tearing of membranes, etc. It is important to avoid pinching the uterus with the forceps. A second, and perhaps related, cause of failure lies apparently in whole inoculum loss, since recipients tend to have either no embryos implanting or several. This can perhaps best be avoided by making the puncture hole no larger than necessary, by using as small a pipette as possible, and by passing the aperture into the lumen for several millimeters.

3. Transfer of Cleavage Stage Ova Into the Tubal Ampulla

Tarkowski (96) was the first to report tubal transfer of mouse eggs, but he did not describe the method. Later, Noyes and Dickman (82) developed a well-described method for the rat, and Whittingham (113-14) applied this to the mouse. Whittingham's method is given below.

With respect to tubal transfer, timing of pregnancy or pseudopregnancy is less critical than in the case of transfer to the uterine horns, although the procedure is more difficult. Tarkowski found that two-cell ova may be transferred with a high rate of success to recipients mated one or two days *after* ovulation of the donor; Whittingham found that any cleavage stage to morula may be transferred successfully into the tubal ampulla of half-day pseudopregnant recipients.

Because ova to be transferred are injected through the tubal ostium into the ampulla it is strongly recommended that one first become proficient at flushing of oviducts (p. 29). The flushing process, in which the oviducts are removed to a dish, is the most convenient and effective way of appreciating the anatomy of the oviducts and especially of locating the ostium. Further, entering the ostium is easier when a modified 30-ga. syringe needle is used, as in flushing (p. 29). Foster transfer by this route, however, necessitates entry with a micropipette, which is more difficult.

Although ampullary transfer may be performed by a single operator, two are recommended, one acting to anesthetize and incise the recipients and to apply wound clips after the transfer, and the other performing the actual transfer. The recipient mouse to be prepared for transfer must be affixed securely to a small board which may be placed on the stage of the dissecting microscope and moved freely. Incisions are made laterally, just below the lowest rib. Usually only one incision is made, since uni-lateral transfer provides a useful control in the opposite oviduct.

Strong incident illumination must be used for the trans-fer, and two dissecting microscopes—the second for load-ing ova—are preferable. These should be close together so that the operator can load the micropipette at one and then swing around to the other, under which the recipient mouse is affixed, holding the loaded pipette in his hand, without changing seats.

A supply of very finely drawn micropipettes is prepared beforehand, and each is compared with the ova to be transferred to be sure that the pipette diameter is as small as possible without causing fragmentation or distortion of the ova. A culture dish is placed upon one microscope, and the illumination is adjusted so that the ova are readily visible in microdrops. A recipient mouse is placed on the stage of the other microscope, and the ovary, oviduct, and a portion of the uterine horn are drawn through a slitted gauze pad. The ostium is then located with watch-maker forceps, and the capsular membrane, if covering the area occupied by the ostium, is torn or pushed aside to provide access. The mouse should then be oriented so that the ostium is conveniently positioned for entry by the operator's micropipette. A pipette is picked up and loaded with several ova. The operator swings back to the recipient and picks up a needle or watchmaker forceps with the other hand. The heels of both hands are firmly steadied by resting them on the supports, and the opera-tor, while looking through the microscope, positions the tip of the micropipette at the ostium. The needle or for-ceps held in the other hand is used to position and hold the terminal tubal segment, and the pipette is inserted in the ostium so that its tip is visible well within the ampulla. The ova are ejected into the ampulla in a small volume. Although the tightness of fit at the ostium will prevent the fluid from flowing back out while the pipette is in posi-tion, fluid will be lost (and often ova as well) when the pipette is withdrawn unless a suitably small fluid volume is used. For withdrawal, the needle or forceps are used to fix the ostium and the pipette removed. Because of the near transparency of the ampullary segment, the operator may often see the deposited ova within it and confirm placement.

It will be found that the most difficult part of this procedure is to enter the ostium with the micropipette without lacerating it or other portions of the oviduct; even with suitably small micropipettes the ostium is quite a tight fit. Furthermore, the pipette tip is not beveled and the edge is quite sharp, making smooth entry more difficult. Probably a better procedure is to expand the ostium by penetrating it beforehand with a rounded, honed 30-ga. needle (p. 29) as is used for flushing.

The incision in the operated recipient is closed with wound clips.

4. Transfer of Oocytes and Zygotes Beneath the Ovarian Capsule

Runner, et al. (88-90) have shown that uncleaved ova may be transferred beneath the ovarian capsule of females of the same stage of pseudopregnancy (zero to one-half days), where they are carried into the oviduct naturally. Of ova so placed, 14% were represented as viable fetuses 19 days later. This experiment may also be performed by using unfertilized ova released by the administration of gonadotrophins and transferring them to the capsule of unmated recipients. These may then be mated immediately after the operation, if hormonally primed. The transferred ova become fertilized and implant normally.

Preparation is the same as for ampullary transfer (p. 46); again, two operators are desirable. The oviduct is exteriorized by gripping the ovarian fat pad with watchmaker forceps and pulling it through the incision until the oviduct is exposed. The operator then loads a capillary pipette with the desired number of ova. Using the dissecting microscope at low magnification, he forces the tip of the pipette into the fat pad, through the capsular membrane, and into the bursa proper (Fig. 21). In this position the tip of the pipette is clearly seen, even though a dissecting microscope is not used. The ova are expelled in a small volume and the pipette is withdrawn, after which the ova should remain visible within the capsule. To close the animal, wound clips are used on the skin; the muscular incision is left unsutured.

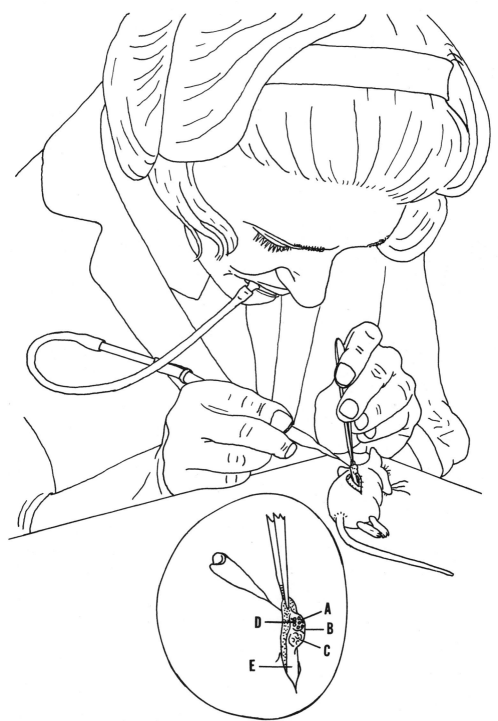

Figure 21. Transfer of One-to Two-Cell Ova beneath the Ovarian Capsule. The exteriorized ovary, oviduct, and uterus (usually drawn out through a slit gauze square) are put on a stretch by traction on the fat body. Micropipette is inserted through the fat body from behind *(inset)* so that the tip appears between ovary and bursa. Ova are expelled into the bursal sac and pipette withdrawn. A. Ovary; B. Ovarian bursa; C. Oviduct; D. Tip of capillary pipette; E. Uterus. Fat body is stippled.

5. Assessing Results

Although transfer success rates are often constant in various laboratories, they are not usually very high. A success rate of 20% or more, measured in terms of term fetuses obtained versus ova transferred, is generally regarded as satisfactory. Because of the constancy of results obtained in many laboratories under a variety of experimental conditions, it is felt that the failures probably do not represent defective ova for the most part. In addition, in a few laboratories in which transfers are frequently made and in which a great deal of experience is a factor, the success rate may be substantially higher, up to 50%.

Experimentally, the assessment of results nearly always must be expressed in terms of viable term fetuses, since those which die in gestation may do so as a result of the experimental procedure. Moreover, if pregnancy is interrupted deliberately, the control series may fail to indicate whether death or deformity might have occurred later in the experimental group. On the other hand, it is inadvisable to allow birth to occur unattended since the females may eat the young. The usual procedure, therefore, is to terminate pregnancy just before the expected time of parturition and to assess the results at that time. In most laboratories, autopsy of recipient females is performed on day 19 in terms of the original ovulation. Since time in culture must be accounted for, mice receiving blastocysts, for example, may be autopsied 16 days after transfer, etc.

The effectiveness of vasectomy of males used to induce pseudopregnancy must always be controlled. One method nearly always employed is unilateral transfer, since in the mouse (unlike some other mammals) ova from one horn do not seem to move across to the other. Control by means of genetic markers has been discussed.

I. THE TRANSPLANTATION OF BLASTOCYSTS TO ECTOPIC SITES

A curious fact is that blastocysts implant readily in a variety of ectopic sites, without regard to the sex or physiological condition of the host. It is especially interesting because the uterus rejects blastocysts except when it is primed in estrus, and so rejection may be an important function of the endometrium.

Ova have been transplanted to the anterior chamber of the eye (34, 87) and the capsule of the kidney (33, 51), to the spleen, scrotal and cryptorchid testis, and peritoneal cavity (9, 34, 54-55, 64). The general finding has

been that either trophoblast or inner cell mass may develop at any of these sites, but that usually one or the other predominates. It is generally thought that overgrowth of trophoblast, the most frequent result, impairs differentiation and growth of the inner cell mass. On the other hand, differentiation and growth of the latter is not commonly striking, even when it does occur. That the uterus may be important specifically in development of the inner cell mass, or embryonic precursor proper, is suggested by the results of a student (S. Michigan, personal communication) who observed enhanced development of both the inner cell mass and trophoblast when endometrial fragments were transplanted beneath the kidney capsule along with ova. Non-trophoblastic differentiated structures which could be recognized were: yolk sac, amnion, brain and neural tube, capillaries and blood elements, epithelial vesicles resembling nephrons, heart, mesoderm, cartilage, somites, and pharynx. In most instances, however, embryonic organization was poor. Curiously, Billington, et al. (10) have found that ova which are cultured to the blastocyst stage and then transplanted to kidney or testis give much better embryonic development than do blastocysts which developed *in vivo*.

A method for transplanting ova to the renal capsule, derived from Fawcett (33) and from Kirby (51), is given below. It is also applicable to the peritoneum and to the capsule of the spleen.

Fortunately, the problem of transplanting ova to ectopic sites and then recovering them later for study is simplified by the usual finding that the trophoblast enlarges somewhat and induces conspicuous local extravasation of maternal blood, forming a hematoma-like nodule which may be 2-3 mm. across.

Two operators are almost essential for subcapsular transplants. One should collect and wash the blastocysts, for which purpose superovulated and mated mice are prepared beforehand. The other operator acts as surgeon.

A 2-3 mm. longitudinal incision is made in the dorsal midline skin, extending caudally from the level of the last rib. Because the mouse's skin is loosely bound at this point, the incision may be moved to the right or left to give surgical access to each kidney in turn. The kidneys may be palpated immediately below the last rib on either side, and the white renal fat is often visible through the thin

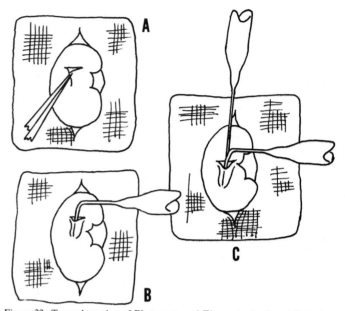

Figure 22. Transplantation of Blastocysts and Tissues to the Renal Capsule. Kidney is exteriorized through gauze square. Renal capsule is grasped with watchmaker's forceps and torn slightly (A). Blunt glass rod is inserted in the opening and worked to produce a pocket (B). Ova are inserted by the second operator with a capillary pipette while the pocket is held open (C). Tissue fragments may be pushed in before or after with watchmaker's forceps. If carefully done, operation is bloodless.

abdominal musculature. A lateral incision is made through the musculature on one side and the kidney presented. It is passed through a slitted gauze pad which, in this case, serves the additional purpose of preventing retraction (Fig. 22). Before the operation, two or three glass rods are prepared, for the purpose of separating the capsule in order to make a pocket to receive the ovum and/or other tissues which may be included. The method for making the glass rods is illustrated in Figure 23. A capillary pipette is drawn in the usual way (p. 15) but is made some-

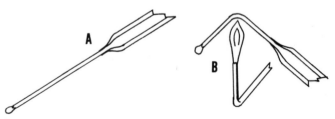

Figure 23. Preparation of Glass Rods for Renal Capsular Transplants. A capillary pipette is drawn in the usual way (Fig. 7) and the end sealed in a flame (A). It is then held briefly in a microflame to produce a right-angle bend.

what larger in diameter by using a shorter pull; after breaking, the capillary should be about 0.2 mm. in diameter. The end of the capillary is then sealed in the microburner, and the closed tube thus formed is bent at right angles, about 0.5 cm. from the tip.

Activities of the two operators must be coordinated. As the surgeon prepares to make the capsular pocket, the other operator, who will inject the ova, should be prepared to pick up the desired number of blastocysts on short notice. Using two pairs of watchmaker forceps, the surgeon grips the capsule at points close together, and, by pulling the tips of the forceps apart, makes a small tear (Fig. 22). The bent and blunted rod is then worked into the tear, parallel to the renal surface, and a small pocket of the capsule lifted off. When the pocket is ready, the surgeon signals the other operator to load the capillary pipette with ova and orients the mouse so that it is in a convenient position for the "ovum operator" to insert the pipette in the pocket without excessive change in position. The pocket is held open with the rod and the pipette containing the blastocyst or blastocysts is inserted beneath the rod or alongside it. When the pipette is in place, the rod is slid out and removed. The surgeon now closely watches the top of the capillary column in the pipette while the other operator expels the ova. Since these should be close together at the tip of the capillary, only a very short movement of the column is necessary to expel them. When such a movement occurs, the surgeon signals his partner who withdraws the capillary.

If other organ fragments, such as endometrium, are to be introduced along with the transplanted ova, they may be added before or after. The best practice, however, is to add them beforehand, since the fragment will be large enough to observe through the capsule and is not readily displaced. If it is added after the ova, however, some or all of them may be forced out of the pocket without the surgeon's knowledge.

Tissue fragments for co-transplant should, of course, be prepared beforehand. Additionally, as a check upon the number of ova actually deposited in the capsular pocket, the surgeon's partner should flush out the pipette in a microdrop and examine the drop for remaining ova.

An advantage of the dorsal midline skin incision is that it minimizes the chance of sepsis post-operatively, since

the various openings do not coincide. In fact, it is unnecessary to sew or slip the abdominal musculature under
these circumstances; two wound clips in the skin incision
are sufficient.

In the performance of an experiment one must decide
when to autopsy the transplant hosts, taking into account
the time required for differentiation and the fact that past
a point of development the transplants begin to regress.
Autopsy is usually done about five days after transplantation, or eight days fertile age. Regression begins at about
day 10 (fertile age), or sooner, in the case of inner cell
mass derivatives; trophoblastic growths may persist until
day 15 or longer, but they apparently always disappear by
the expected parturition time.

The interpretation of sectioned autopsy material may
be difficult, owing to the complex "inverted" character
of mouse development, and to the fact that the orientation
of any embryonic mass present cannot be determined
beforehand, and so the plane of section is random. If the
experimenter is reasonably familiar with the peculiarities
of rodent development, it is not exceptionally difficult
to identify organs and tissue types (86). It is usually rather
difficult, however, to assess accurately the quality of embryonic organization. Needless to say, the entire transplant should be sectioned serially.

J. THE FORMATION OF MOSAIC
EMBRYOS BY FUSION OF OVA

When mouse eggs are fused, they may be cultured to the
blastocyst stage and, if desired, transferred to foster females to obtain mosaic fetuses (67, 71-72, 98-101). Because of the two-cell block, this stage is the earliest that
can be utilized if the resulting mosaics are to be grown in
culture. Fusion results in apparently random intermingling
of cells in the resulting embryo, with the qualification that
if fusion is carried out late in cleavage (blastocyst or late
morula), labeling studies have shown that intermixing is
only partial, presumably because of the number of cells
and the amount of time left before embryonic differentiation occurs (67, 71-73). In addition, the fusion of
blastocysts does not succeed well or yield a very high per-
centage of viable fetuses, and in fact there is reason to
suspect that determination may begin at the late morula
stage (98-100). In any case, the best results in terms of
normal fetuses occur when stages from two-cell to morula
are fused.

As a preliminary, the zona pellucida must be removed. This is normally accomplished with pronase, taking care to avoid letting the enzyme act upon the egg surface for any time. Tarkowski allows digestion to proceed until the zona is very thin, then removes the ova to medium and rids them of the remaining zona coat by pipetting with a capillary of suitably small bore (100). Mintz, however, claims that reasonably prolonged pronase treatment of naked ova does not lead to reduced viability of the resulting mosaics (66).

An important consideration in the formation of mosaics is stickiness of the blastomeres. Although stickiness is obviously necessary for fusion to occur, it introduces the complication that denuded ova tend also to stick to the pipette and the culture dish. Fortunately, however, denuded eggs are much more cohesive than adhesive. Their stickiness is controlled by two factors: the presence of serum and high temperature. If medium containing 50% fetal calf serum is used, and arrangements are made to allow the initial contact to take place at 37° C., permanent contact occurs within a minute or so, and the resulting mosaics may then be returned to the incubator with little or no risk of separation (72). However, it is recommended that culture dishes and pipettes be siliconized.

In foster transfer of fusion mosaics no special precautions are necessary beyond the obvious need for gentle handling, since the zona is no longer present to provide support and protection. Foster transfers are usually made with mosaics that have progressed to the blastocyst stage, in which case transfer is made directly to the uterus (p. 42). If earlier stages are to be transferred, they may be placed within the tubal ampulla (p. 46).

A final comment concerning terminology may be worthwhile since fusion products have been referred to as "mosaics," "chimeras," and "allophenes." The last of these terms was introduced by Mintz (71, 73) and denotes that the phenotype of fusion embryos is a function of the phenotypes of persisting donor cells. The term "chimera" may be objected to on the ground of its denoting monstrousness, whereas fusion embryos are often quite normal and even regulate in terms of cell number and size. The term "mosaic" is suitable but would not apply in the case of fusion products in which the cells of a fusion egg have a lethal genotype and later disappear.

1. Removal of the Zona
Pellucida with Pronase

The zona pellucida is readily removed with a 0.5% solution of pronase (as Calbiochem) dissolved in a balanced salt solution such as Hanks's BSS or Ringer's Solution (43-44). Digestion proceeds well at room temperature, and, since bicarbonate is not needed, it is usually omitted, obviating the need for ambient gassing or for incubation.

The pronase solution may be made up in stock batches and filter-sterilized. If stored at 4° C., the solution retains its activity at a useful level for a week or so. In our experience, frozen solutions may be kept for quite a long time without loss of activity (at least six months at −15° C.), and in fact a stock batch may be thawed for use and refrozen several times, although a precipitate may form after refreezing. Perhaps the best practice is to store the enzyme solution frozen in 1.0 ml. aliquots, so that each aliquot may be discarded after it is thawed and used.

Before beginning digestion the work should be well planned and dishes set up with microdrops or other culture vessels made ready and equilibrated beforehand. Ova are transferred to small amounts of the enzyme solution, usually in microdrops or layers under oil, and observed continuously with the dissecting microscope. Digestion should be complete in three to five minutes, earlier stages usually requiring the longer time. As digestion proceeds, the zona will be seen to begin thinning rather suddenly; soon afterward it disappears completely. The process may be aided by gently pipetting the ova as digestion proceeds. As noted, some workers (i.e., Tarkowski, 101) prefer to stop the process before digestion is complete in order to avoid possible deleterious effects of the enzyme acting directly upon the surface of the blastomeres.

When the zona is removed, the ova are washed three to five times in Hanks' or Ringer's Solution and then transferred to a medium appropriate for the experiment to be carried out. With the exercise of reasonable caution in pipetting, the blastomeres do not separate.

2. Fusion of Ova

a. The culture dish and
watchglass methods

As noted, naked ova are quite sticky in the presence of serum and fuse spontaneously once contact is made if the medium is warmed to 37° C. It should be noted, however, that the blastomeres may stick to glass as well as to each other, though the force of adhesion is fortunately much less than the cohesive affinities. Care must be exercised when naked ova are pipetted in serum-containing media to avoid adherence to the inside of the pipette. Siliconized pipettes and culture dishes may be used, or pipetting may be done fairly rapidly so that ova have little opportunity to form attachments to the glass. The exercise should be arranged so that most of the handling is accomplished in Hanks' or Ringer's Solution. If a relatively large volume of medium is used (0.5 ml.), the naked ova, provided they are handled in small volume, may be transferred to it for fusion without need for an intermediate wash in medium. Tarkowski (98 - 100) succeeds in fusing ova using Mulnard's medium (see appendix), containing bovine serum albumin at a concentration of $3 - 5\mu$ g/ml., instead of serum. Denuded ova are placed in a microdrop and forced into contact by removal of most of the medium. After a few hours, incubation medium is added to re-form microdrops.

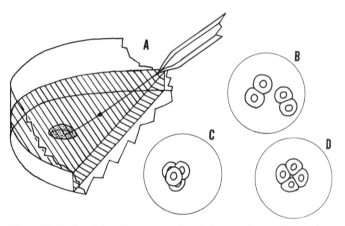

Figure 24. Fusion of Ova. Two or more denuded ova are inserted into a microdrop of medium under oil (A). The contents of the dish are warmed and the ova within a drop (B) are pushed together (C). After reincubation for a few hours the blastomeres organize into a mosaic embryo (D).

Fusion may be accomplished either in microdrops in culture dishes or in embryological watch glasses, using Mintz's medium (72; see appendix). In either case, an oil layer is often used. In the first instance, two or more ova are transferred to a warm microdrop following an intermediate wash in medium. Fusion is accomplished by pushing the ova into contact with each other, using a fine needle or even the tip of the micropipette (Fig. 24). Adherence should occur immediately, but it is good practice to leave the adherent ova undisturbed for five to ten minutes before they are transferred to the incubator. During subsequent incubation, intermixing occurs, accompanied by rounding of the fusion product to give a large but otherwise normal embryo appropriate to the stages fused. Although intermixing appears to be random, as previously noted, it may not be complete if multicelled donor ova are used, presumably because the positions of blastomeres become fixed before there is time for intermixing to result in complete randomization of cells. Thus, when unlabeled and thymidine-labeled blastocysts are fused and then transferred to foster recipients, the resulting embryos often show an unequal distribution of labeled nuclei but with many labeled nuclei widely scattered.

Foster transfer of allophenes succeeds about as well as does transfer of unfused ova with zonas intact, provided that suitable care is exercised in manipulation after zona removal and that the route of transfer and pseudopregnant age of the foster recipient are appropriate for the developmental stage being transferred (p. 39).

b. The hanging drop method

Naked ova may also be cultured for a time in inverted droplets. In this case the ova fall to the bottom of the droplet and are brought into firm contact automatically. They may then be transferred to microdrops or to oil-layered medium after several hours incubation. In using this method it is important to take precautions against drying of the droplets.

Plastic culture dishes are used containing several milliliters of distilled water or balanced salt solution in the bottom part. The top part is removed and inverted. One to

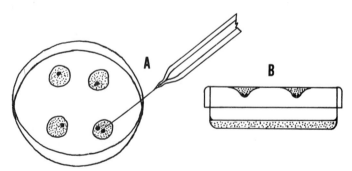

Figure 25. Hanging Drop Culture for Ovum Fusion. Microdrops are set on inverted lid of disposable petri dish (A), and denuded eggs are added. They are pushed as close together as possible, and the dish is inverted (B) over the bottom portion, which contains medium or saline to prevent drying. During incubation the ova settle to the point of the microdrop and are forced into contact.

three microdrops are spotted on the top part, and ova for fusion are added to them (Fig. 25). The top is then raised and inverted quickly with a flip of the wrist so that the drops are now allowed to run. The top with the adherent hanging drops is placed back over the bottom and the unit placed in a glassed, humidified incubator. After a few hours the fused eggs may be transferred to microdrops under oil.

3. Formation of Mosaics by Injection

Gardner (37) has produced mosaics by injecting separated blastomeres into the blastocyst cavity, which has the theoretical advantage that unbalanced mosaics may be produced with a small number of "donor" cells and a relatively large number of "host" cells. A further advantage is that the method should allow more complete intermixing of cells than occurs when whole blastocysts are fused. In Gardner's experiments, donor cells bearing a pigment gene marker and the T6 chromosomal translocation marker were used. Chromosomal mosaicism in fetal tissues was noted, as well as iris and coat color mosaicism in newborns. An interesting finding is that extensive mosaicism occurred after injection of only three cells, suggesting that the embryonic progenitor cells of the blastocyst are very few in number.

An operational difficulty arises from the fact that the procedure is difficult and requires experience. Blastocysts are held with a suction pipette in a micromanipulator, and the relatively thin zona is torn at one point with microneedles. An inoculating micropipette of very small bore, containing one to five cells, is inserted into the blastocyst cavity, and the cells are injected using a micrometer syringe. Successfully injected blastocysts rapidly reheal and are transferred to foster females, by intrauterine injection (p. 42) for further development.

K. THE SEPARATION AND CULTURE OF BLASTOMERES

The question of regulative versus mosaic determination in mammalian embryos and the stage at which determination of cell type occurs is of considerable interest. Blastomeres of four-cell and eight-cell embryos can readily be separated and cultured to yield smaller but sometimes normal-appearing embryos (102). These in turn may be transferred to foster recipients with some success in producing viable offspring. However, the question of whether early determination occurs in the mouse has turned out

to be difficult to settle, since advanced cleavage stages
derived from separated blastomeres and subsequent culture consist of abnormal-appearing forms in a high proportion of cases, often lacking the inner cell mass or a
separate trophoblast. Thus, the results suggest that trophoblast may be determined as early as the four-cell stage
(101 - 2), a possibility consistent with the fact that trophoblast differentiation takes place before gastrulation in
the inner cell mass and far removed from it. Blastomeres
from the two-cell rabbit ovum can form apparently normal
embryos, however (91). Nevertheless, the point is by no
means settled, and analysis of blastomere development
remains of great interest for this as well as for other
experimental approaches to development.

It will be clear from Mintz's findings (72), discussed in
the previous section, that blastomere separation requires
procedures that diminish stickiness. These are absence of
protein from the medium, and lowered temperature
(20° - 25° C.). Under these conditions, ova or separated
blastomeres may be fused with great difficulty but do not
fuse upon casual contact and do not ordinarily stick to
glass or to the culture dish surface.

Ova of two-and-a-half days ovulation age are collected
in Ringer's Solution containing 0.1% bovine serum albumin, fraction V (Armour). They are then transferred to
0.5% pronase (Calbiochem) in Ringer's Solution for zona
removal. Because a high proportion of abnormal forms
results from culture of separated blastomeres, the practice
is to remove the thinned zonas by pipetting, in order to
avoid possible damage to blastomeres if they are permitted to contact the enzyme solution (101 - 2). Therefore,
when the outer membrane becomes quite thin, the ova
are removed to Ringer's Solution and pipetted. In practice, this point may be missed, since the zona disappears
very rapidly once the process begins.

Blastomere separation requires the absence of divalent
cations and of serum. Naked ova are transferred to a mixture of Dulbecco's Solution (less Ca^{++} and Mg^{++}) and
Versene (disodium ethylenediaminetetraacetate) to 0.02%
and left to stand at room temperature for 20 to 30 minutes
of continuous observation, during which blastomere separation occurs. The process is assisted by occasional
gentle pipetting. When separation is complete, blasto-

meres are transferred to nutrient medium (see appendix) in microdrops or a medium layer, in either case with an overlying layer of equilibrated paraffin oil. Several blastomeres may be transferred together to the same microdrop or medium layer without fear of refusion, since the medium, although it supports cleavage quite well, lacks serum.

Mulnard (77) removes the zona mechanically by means of fine glass needles and a 40 μ internal diameter suction pipette for ovum immobilization, all held in a Leitz mechanical micromanipulator.

Zona removal is accomplished in Ca^{++}-Mg^{++}-free phosphate-buffered saline at pH 7.3, and containing versene at 10^{-4} g/l. Eggs are sucked gently in and out of a micropipette of 70 μ internal diameter until separated. If cultured in Mulnard's medium (see appendix) the culture dish is precoated with an agar layer to prevent sticking to the dish surface, since this medium contains serum.

To date, the method has consistently yielded a proportion of "false blastocysts" (cell masses lacking a cavity) and of trophoblastic vesicles (lacking an inner cell mass), as well as normal-appearing blastocysts. Figure 26 reproduces illustrations of Tarkowski and Wroblewska (102) of normal and abnormal blastomere-derived ova.

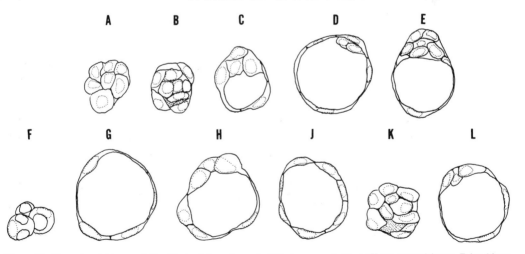

Figure 26. Normal and Abnormal Ova Derived from Separated Blastomeres. A - E and K are normal forms. D is a blastocyst with only two cells in the inner cell mass, derived from one blastomere of a four-cell ovum. The two cells are recognized as belonging to the inner cell mass because they do not contact the surface. F represents a "non-integrated" form derived from one of eight blastomeres. Vacuoles appear in the cells. G and J are trophoblastic vesicles, lacking inner cell mass cells. H and L are "false blastocysts" in that the protoplasmic mass is eccentrically distributed, but all cells contact the surface. A, D, J and L were each derived from one blastomere of different four-cell eggs; the remainder developed from blastomeres of eight-cell eggs. Redrawn after Tarkowski and Wroblewska (102).

L. BLASTOMERE DESTRUCTION

Tarkowski (97) and Seidel (91) were the first to apply the famous Spemann experiment to the mammals, by destroying one blastomere at the two-cell stage and following subsequent development. More recently, Mulnard has continued and extended these studies by following the development of acid phosphatase activity in resulting morulae or blastocysts and comparing the effects of both destruction and separation upon development of the other blastomere (77). In normal development, acid phosphatase activity is limited to the inner cell mass, and Mulnard used this effect to distinguish tentatively between resulting inner cell mass versus trophoblastic cells. After destruction or removal of a blastomere at the two-cell stage, the remaining blastomere usually produced a vesicle lacking an inner cell mass and acid phosphatase negative. In a few exceptional instances, however, solid all-positive masses resulted, suggesting that a tendency toward organization of bilateral symmetry may occur in the uncleaved ovum. However, as shown by other work, blastomeres from four-cell and eight-cell ova often fail to form inner cell masses, even when all blastomeres of a particular ovum are successfully cultured in isolation. Hence it is not possible to draw certain conclusions as yet, and so the question is still open; the possibility remains that nonspecific injury or mass effects may be involved.

Using a mechanical micromanipulator, a two-cell egg is held in place by a suction pipette of 40 μ internal diameter. The zona pellucida and one blastomere are punctured with a finely drawn glass needle, and destruction is confirmed by continuing observation until cytolysis is evident.

M.   FERTILIZATION *IN VITRO*

Although for some time *in vitro* fertilization in the hamster (117) and in the rabbit (11 - 12, 25) has been successful, until recently such experiments have not yielded satisfactory results in the mouse. Brinster and Biggers (22) have obtained low percentages of cleaved oocytes by incubating egg-containing oviduct fragments with sperm. Recently, however, Whittingham (114) has reported 40 - 50% success in repeated experiments, opening important new avenues of investigation. His method is described below.

The two-cell block phenomenon observed with mouse embryos *in vitro* complicates the analysis of results; that is, although the first cleavage succeeds in culture, permanent arrest normally occurs before the second cleavage. Thus it is more difficult to say whether eggs fertilized *in vitro* are normal. However, Whittingham was able to show that two-cell ova produced after fertilization in culture yield viable young when transferred to foster recipients, as will be described below.

The best results were obtained from the use of F_1 hybrid eggs produced by hormonal superovulation in females of C57Bl × Balb C cross, although eggs derived similarly from Swiss females cleaved in 15-25% of cases, following exposure in culture to sperm. Our experience has been that use of hybrid donors is very important.

Donor females for eggs and for sperm are injected with PMS at about 7:00 P.M. three evenings beforehand and then given HCG (3 I.U. of each preparation) at 8:00 P.M. the night before the experiment is to be performed. On the day of the experiment, females that are to serve as sperm donors are mated at 8:00 A.M., one male being used for each female. Between 9:30 and 10:00 A.M., the females are killed and the sperm suspension removed from the uterine horns. Preincubated embryological watch glasses are used, containing 0.5 ml. of nutrient medium overlain with oil. Each uterine horn is dissected free except for the cervical attachment; the horn is then cut across as far caudally as possible and immediately held over one of the preincubated watchglasses. Often the trauma of cutting causes uterine contraction and expulsion of a drop of sperm suspension. This is allowed to fall into the watch glass and the mixture stirred vigorously at the same time; delay in mixing results in coagulation of the seminal fluid. When a drop of fluid does not fall spontaneously from the cut horn, it is expressed out with a second pair of forceps and mixed as before.

The watch glasses may be quickly examined with the dissecting microscope to determine that large numbers of active sperm are present. They are then removed to the incubator while the eggs are prepared.

In this process, cumulus masses are removed intact at 11:00 A.M., by tearing the ampulla (p. 29), and transferred to previously incubated microdrops. If necessary, follicle cells may be removed by pipetting, but *in vitro* fertilization succeeds perfectly well when intact cumulus masses are used.

To each microdrop containing unfertilized eggs an approximately equal volume of sperm suspension is added.

The microdrops may be inspected to be sure that moderately large numbers of active sperm are present but should be returned to the incubator fairly soon. Sperm counts and attempts to quantitate sperm number are usually not done since these measures usually result in delay and since sperm number does not seem to be a critical factor within rather wide limits.

Incubation is continued for seven hours—Whittingham's results indicate that in some instances fertilization is delayed this long—and eggs transferred to prewarmed microdrops of nutrient medium (or to nutrient medium in prewarmed watch glasses), to remove sperm. The eggs are not allowed to cool substantially before being reincubated, and to avoid this a warming plate next to the microscope is very helpful.

If intact cumulus masses are used it will be noted that the follicle cells detach during the seven hour incubation period, leaving the eggs free. Control cumulus masses should be dispersed with hyaluronidase (p. 32) at this point.

Incubation is continued overnight, and the eggs are inspected for cleavage twenty-four hours after sperm removal. Because of the *in vitro* blockage of cleavage at the two-cell stage, there is no need to inspect the cultures at a later time; in fact, abnormal or fragmentation cleavage may eventually occur. One occasionally also observes normal appearing "first cleavages" in unfertilized eggs, and so it is important to control the experiment by incubating some eggs from each donor in sperm-free medium.

In Whittingham's experiments, the normality of fertilization and first cleavage were assessed by foster transfer to the tubal ampulla of pseudopregnant foster recipients (p. 46). These were Swiss females mated to vasectomized C57B1 males without hormone injections; here, the genes for black pigmentation were not required, but vasectomized C57B1 males were used because the induction of pseudopregnancy in Swiss females appears to be more effective than when Swiss males are used. Since the egg donors were F_1 hybrids between albino and homozygous black, while sperm were derived from albino males, the fetuses produced from the transferred ova should be albino or pigmented in a one-to-one ratio. Because this expectation was realized in fetuses recovered on the eighteenth day postovulation, fertilization appears to have been normal and the vasectomized males to have been sterile.

Because there is no evidence that mouse sperm must be "capacitated" (allowed to remain in the uterus for a time) before they can fertilize eggs, it may be possible to fertilize eggs *in vitro* using sperm collected directly from the vas deferens and epididymis (below). As far as I know this has not yet been attempted since Whittingham's success with uterine sperm.

N. ARTIFICIAL INSEMINATION

This is a rather reliable procedure which is especially useful in studying the effects of prefertilization aging upon eggs and sperm. The method described below was introduced by Dziuk and Runner (30) who found that sperm retain their motility for at least 72 hours and their ability to fertilize eggs for about 13 hours. This and other techniques were used to study egg aging by Runner and Palm (90), Braden and Austin (13), and Marston and Chang (60), whose results indicated that mouse eggs can be fertilized at least 13 hours after estimated ovulation. James Schreiber, a student in my laboratory, recently found an upper limit of 15 hours.

The method takes advantage of the fact that large numbers of sperm are stored in the vas deferens of the mouse, enabling one to obtain a suspension free of seminal vesicle product, and hence one which does not coagulate upon contact with air.

Females are conveniently "cycled" and ovulated, usually with super ovulating doses, by injecting PMS and HCG with ovulation assumed to occur 12 hours after the second (HCG) injection. On the morning of the experiment, male sperm donors are killed and the epididymis and vas deferens removed (see instructions for vasectomy, p. 40). The cut end of the vas is held in a drop of about 0.50 - 0.65 ml. of medium in a watch glass kept at 37° C. on a warming plate. A dense, pencil-like suspension of sperm exudes from the vas if the males have been kept isolated for several days beforehand. The epididymis may be added to the suspension and cut into several pieces which are removed later. The sperm rapidly become motile and form a homogeneous suspension.

Dziuk and Runner used a suspension medium of 9.5% reconstituted dry skim-milk powder boiled for ten minutes and cooled to room temperature; tissue culture medium 199 works as well.

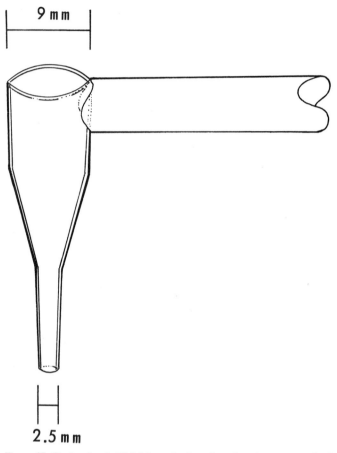

9 mm

2.5 mm

Figure 27. Device for Artificial Insemination. Speculum is constructed of drawn glass tubing to which a handle is fused. After Dziuk and Runner (30).

For insemination, a glass speculum is mounted in a clamp with a focused light shining into it (Fig. 27 and 28). The speculum is flared at one end, the other end consisting of a tubelike projection with an external diameter of about 2.5 mm. and an internal diameter large enough to admit a 22-ga. syringe needle. A small syringe is attached to a 22-ga. needle which has been blunted and bent 90° in such a way that the end projects one-eighth of an inch beyond the small end of the speculum when completely inserted within it.

An ovulated female is firmly grasped and positioned so that the end of the speculum enters the vagina, with the inseminating needle partly withdrawn. By looking down the speculum, the operator observes the cervix and ascertains that it is in contact with the speculum end. The inseminating needle is inserted completely, the operator usually feeling a slight lowering of resistance as it just penetrates the lowest point of the uterine cavity. Then 0.08 to 0.12 ml. of suspension is injected, an amount sufficient to fill both horns with sperm suspension.

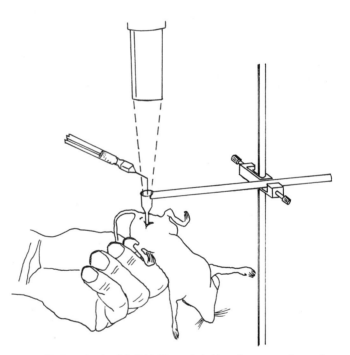

Figure 28. Inseminating Method. Mouse is held so that mounted speculum can be inserted into vagina with the opening lodged against the cervix; the position is checked by looking down the speculum with the aid of the illuminator. A bent blunted needle, of appropriate length and attached to a syringe containing sperm suspension, is inserted as far as possible and the fluid injected into the uterus. Redrawn after Dziuk and Runner (30).

The procedure is easily mastered with a few practice trials. Sperm may be collected and stored for various periods beforehand if sperm aging effects on storage media are to be tested. Within limits, the experiment can be planned for convenient times of day, since PMS and HCG injections may be given females at any time between about 9:00 A.M. and 10:00 P.M., ovulation occuring approximately 12 hours later. Sperm, of course, may be collected at any time.

Sperm survival *in utero* may also be tested by flushing uteri at suitable intervals after mating. Finally, as noted, insemination may be delayed to study the effect of aging on eggs.

Results may be assessed either by flushing tubal or uterine ova and determining the proportion of cleaved ova, or by allowing pregnancy to proceed to a later stage. Especially in the first case, controls should be meticulous, since unfertilized ova sometimes cleave once ("abortive cleavage").

O.   IMPLANTATION *IN VITRO*

The study of implantation of mouse ova in culture represents an almost entirely untapped field, since it has hardly been attempted, and since the mouse ovum differs from the rabbit ovum—with which it has been well studied by Glenister (40)—in a number of aspects which may well have a bearing upon implantation *in vitro*. For example, the rabbit blastocyst becomes greatly expanded, whereas that of the mouse does not. Further, the mouse ovum exhibits the phenomenon of delayed implantation, while the rabbit ovum does not. Either factor may influence implantation *in vitro*.

Because this area seems especially important, some attention will be given to Glenister's methods and results with the rabbit, on the ground that many ought to be applicable to the mouse also.

Successful results have been achieved with organ-type cultures involving either a base of agar overlaid with biological fluids or defined medium (such as medium 199) or, alternatively, the culturing of tissue fragments on teabag paper over a wire mesh platform. Because the latter method is most suitable for studies limited to defined media it will be described here. For further details concerning this and other methods the reader may consult Biggers, et al. (5, 7), and Jensen, et al. (48).

Various grid- or raft-type culture systems may be used, but perhaps the most satisfactory from the standpoint of simplicity and convenience is the Falcon organ culture dish (#3010) and organ culture grid (#3014). Culture dishes are prepared by placing a grid over the center well and adding to the well enough medium 199 (about 1 ml.)

to fill the well and the interstices of the grid (Fig. 29). The absorbent paper ring peripheral to the well is thoroughly wetted with about 1 ml. of sterile distilled water in order to retard evaporation from the medium, and the dishes are incubated in a gassed chamber to allow equilibration. If uterine tissue is to be used, fragments of whole uterus measuring a few millimeters on a side are removed with sharp knives or scalpels; scissors cause crushing and should be avoided. The fragments are transferred to medium or to saline and the myometrial and mesometrial layers removed and discarded, leaving the endometrial layer. These are placed upon grids (as many as three per grid) and the cultures reincubated. Six-day rabbit blastocysts are then collected (development to the time of implantation requiring a longer time in the rabbit than in the

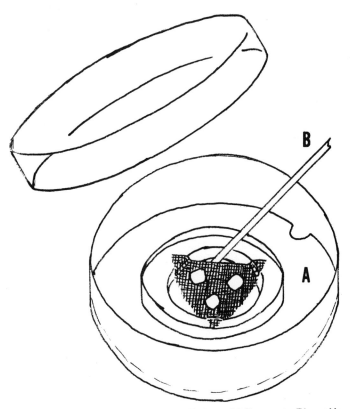

Figure 29. Setting Grid Cultures with Endometrial Fragments. Disposable organ culture dish (Falcon) is illustrated. Filter paper ring (A) is saturated with sterile distilled water or saline. Stainless steel mesh grid is put into place and medium added with a Pasteur pipette (B) until capillary contact is made with the mesh. Endometrial or other tissue fragments are placed on the grid and blastocysts may be pipetted on top of the fragments.

mouse) and the zona pellucida removed with watchmaker or other fine forceps, a relatively easy procedure in this case. One to several blastocysts are then placed upon each fragment, either on the endometrial or mesometrial surface, and the cultures are returned to the incubator. Degeneration often begins after about six additional days, so the cultures are usually harvested for study at that time.

The addition of serum or other biological fluids to the medium does not appear to enhance implantation, but high oxygen content not only increases the incidence of implantation but also prolongs survival time. Hence the gassing mixture commonly used consists of 95% oxygen and 5% CO_2. Under these conditions, some three-fourths of the blastocysts implant, and a high proportion of those accomplish the striking process of blastocyst expansion which is so characteristic of the rabbit ovum.*

Unfortunately, however, there is some question, indicated by Glenister, as to what degree the process seen in culture mimics that seen in the intact animal. Although blastocyst expansion appears to occur normally, and differentiated structures such as heart and neural tube appear, an organized embryo does not result. Similarly, implantation occurs as well upon either the endometrial or mesometrial surface and upon the chick chorioallantoic membrane. In addition, it does not seem to matter whether uterus from a virgin doe, isologous uterus, or uterus from a doe in the same stage of gestation as the egg donor is used. Thus, priming for a decidual response clearly is not crucial, although some response does seem to be induced by the ovum upon contact with uterine fragments. On the other hand, histological and ultrastructural study indicates that the invasion differs in some respects from that seen *in vivo*.

With respect to development of the embryo proper, much evidence indicates that, while high oxygen tension is required for development in culture after implantation has occurred, it is deleterious to preimplantation embryos such as blastocysts. Glenister believes that this may explain the occurrence of differentiated but unorganized embryos in this system.

*It should be noted that use of gas mixtures high in oxygen content is hazardous: safety precautions should be investigated beforehand.

P. THE CULTURE OF
IMPLANTED EMBRYOS

Early somite and even presomite embryos (8 to 9 days) may be explanted to plasma clots or homologous serum, after which they form another 8 somites about every 12 hours for the next 40 hours, when they begin to deteriorate at the 30-35 somite stage. At this point, the embryos have a functioning blood circulation, prominent anterior and posterior limb buds, and appear normal in most or all respects. One hardly knows where it will all end.

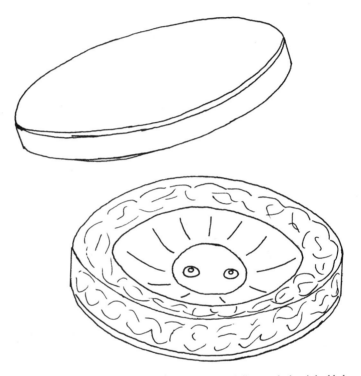

Figure 30. Watch Glass Culture with Plasma Clot. A flat watch glass is bedded on wetted cotton in a petri dish. Plasma (15 drops) is added, followed by embryo extract (5 drops). After clot formation, embryos are placed on the clot surface and a drop of medium added, smoothing it over the embryo surface so that the entire embryo is wetted. Development may continue for 40 additional hours, from pre-somite to 35 somite stage, if incubated in high oxygen environment. After New (78).

The most successful methods have been developed by Smith (92), and by New (78-79) and New and Stein (80-81). They involve the culturing of early somite embryos on agar, plasma clots, or in homologous serum, in an atmosphere rich in oxygen. The method of New and Stein is described.

Cultures are set in watch glasses within petri dishes containing cotton or absorbent paper at the margin (Fig.

30); it is well wetted to retard drying. To each watch glass, chicken plasma (15 drops) and embryo extract (chick, rat or mouse, 5 drops) are added and mixed. Clotting should occur within ten minutes. The beginner should be cautioned, however, to be certain beforehand that the plasma does indeed clot in the presence of embryo extract; commercial plasmas are heparinized or citrated and will not clot without extensive dialysis. Methods for preparing both plasma and embryo extract are given by New (78). Embryo extracts of chick, rat, and mouse origin seem to be equally efficacious, and later experiments indicated that explanted embryos grow as well in homologous serum in watch glasses as on clots.

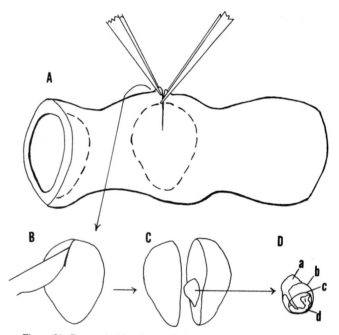

Figure 31. Removal of Implanted Embryos for Culturing. Uterus is grasped with forceps at region of conceptual swelling and torn (A). Deciduoma is removed intact and incised shallowly around its long axis (B). The halves are separated (C), and the embryo (D) is removed with membranes intact. a, ectoplacental cone; b, yolk sac; c, amnion; d, embryo. After New (78).

Females of eight to nine days pregnancy are killed by cervical dislocation and the uteri removed aseptically to a balanced saline such as Tyrode's solution. The massive deciduomas, each containing a conceptus embedded within, are then removed by tearing open the uterine wall with watchmaker forceps. With care, the pear-shaped deciduomas may be removed intact and undamaged (Fig. 31). Using the dissecting microscope, the embryos, with their membranes intact, are then dissected free from the deciduomas. With a sharp scalpel, a longitudinal incision is made all around the deciduoma without penetrating deeply enough to reach the embryo, which is located nearer the acute end of the mass. Forceps are then used to complete the separation of the two halves. The embryo generally remains attached to one of the halves, from which it is freed.

At this point, because of the peculiar inverted development of rodents, the yolk sac surrounds the embryo, with Reichert's membrane external to that and concentrated at one pole; at the other pole, the ectoplacental cone protrudes (Fig. 31). Reichert's membrane should now be grasped with very fine watchmaker forceps and torn away, leaving the yolk sac essentially denuded.

Returning to the culture clots, a drop of embryo extract is placed on the top of each clot and an embryo added, with the yolk sac intact. It is oriented so that the ectoplacental cone is to one side, and the drop of embryo extract is spread out with the tips of the forceps until its surface is just level with the top of the explant.

The embryos are incubated at 37° C., in an atmosphere of 5% CO_2, 60% O_2, and 35% N_2.

In setting up the cultures, it is not necessary to hurry in removing the embryos to the hyperoxygenated environment, since a three-hour delay after killing of pregnant females does not seem to affect subsequent growth and development of the embryos. Further, New reports (79) that only five to ten minutes per embryo are required to set cultures. In removing Reichert's membrane the yolk sac is occasionally nicked, but such nicks usually heal and development occurs normally. Smith (92) removed the yolk sac altogether, exposing the embryo and making it accessible to short-term study of the effects of extirpation, etc. Although such embryos evidently do not thrive as well as when the yolk sac is intact, the method is promising as a technique for longer term experiments.

Embryos older than eight to nine days may also be explanted, and in fact they give higher survival rates. In no case, however, has development proceeded much beyond about 35 somites. This may be due to limits upon the efficacy of gaseous exchange, since development to this point is quite dependent upon high oxygen concentration. The presence of CO_2 in the atmosphere, while less critical, seems also to help.

Q. THE RECOVERY AND PREPARATION OF STAGED EMBRYOS

In the majority of experimental situations, data are recorded directly from fresh material, or from photographs of fresh material, which is discarded. In some cases, however, there is a need for permanent or semipermanent preparations.

1. Oocytes to Blastocysts

Because of the small size of preimplantation stages, it is not feasible to use standard cytological methods, which require a great deal of manipulation in transferring to different solutions, etc. The most practicable method of managing early embryos seems to be to attach them firmly to slides and to manipulate the slides in solution changes.

A few general points should be made, however. One is that all of the methods used cause some distortion of ova and should be avoided when this is a factor. The fixative used is also extremely important in controlling distortion. In particular, alcohol or alcohol-water solutions should never be used as the primary fixatives for any embryos since they cause drastic shrinkage. When relatively undistorted ova are needed for records or study, perhaps the best method is to fix them in osmium tetroxide (2% aqueous solution for four hours) or in 10% formalin (12 hours or longer). Glutaraldehyde fixation is also excellent. The ova can then be transferred to 70% alcohol for study or permanent storage. Such ova must, of course, be manipulated by micropipette. They should be stored in a small volume of fluid in small containers so that they may be conveniently recovered.

One of the most satisfactory methods for attaching ova to slides is that of Austin (2), in which a coverslip is supported at the corners in such a way as to press lightly upon the ova and hold them in place while solutions are added and withdrawn by capillarity. The method is described on page 22. A disadvantage of the method is that it is only applicable to oocytes and uncleaved or two-cell embryos, since the coverslip pressure causes disorganization of ova in later cleavage stages. A second disadvantage is that it is rather difficult to make permanent preparations with the method, although many preparations may be stored for several months in the refrigerator. Finally, when acetic acid solutions are used in this or any other method, the zona pellucida dissolves (43), a factor which may or may not be important in particular cases.

A second method for making preparations of early stages is to affix them to slides with albumin, which has the advantage that permanent slides are easier to prepare. On the other hand, the method requires partial drying and so it is impossible to prevent collapse of later cleavage stages, especially blastocysts. The method used by Porturas (personal communication) is given below.

Eggs are recovered by flushing in the usual way. Coat microscope slides by adding a drop of albumin and spreading with a finger, wiping off most of the albumin in the process. Add several eggs in a small volume of saline and draw off as much as possible. Let the slides dry almost (but not quite) completely and then place them in a Coplin jar containing a few cc. of 10% formalin, so that the eggs fix in the vapor. Cover the jar and let it stand for thirty minutes or longer. Wash by dipping in tap water (one minute) and in distilled water (a few seconds). Transfer to a solution of 0.25% Harris' hematoxylin for one to two minutes, then to 3% lithium carbonate. Wash by dipping in two changes of distilled water and transfer to 1% aqueous eosin Y for one to two minutes. (This step may be omitted if cytoplasmic staining is not desired.) Wash by dipping in tap water and dehydrate by passing rapidly through ascending alcohols (70%, 80%, 95%, 100%). Clear in two changes of xylol (three minutes each). Without allowing the slides to dry, add a few drops of Permount and put on a coverslip, being careful not to crush the ova.

2. Demonstration of Early Implantation Sites

Early implantation sites are difficult to demonstrate by usual staining methods but can be shown in five-and-a-half to eight-day material by clearing and bleaching to take advantage of pronounced vascular changes which occur soon after the blastocyst contacts the endometrium (Orsini, 83). Uterine horns are pinned out on a piece of cork or cardboard (Fig. 32) and fixed in 10% formalin overnight. They are then washed for several minutes in tap water and simultaneously bleached and dehydrated by transferring to 70% alcohol which contains H_2O_2 to 10%. Bleaching is allowed to continue until implantation sites are readily visible, and preferably until the areas between

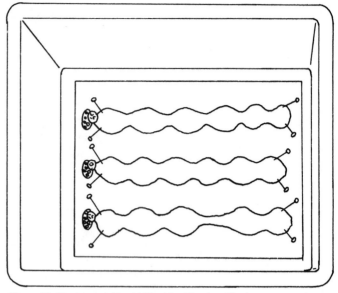

Figure 32. Pinning out Gravid Uteri for Fixation. Uteri are removed intact and pinned out on a thick cork sheet in stretched form. The sheet is placed in a staining dish and fixative or other solutions added.

them are completely white. It is often necessary to change the alcohol-peroxide solution several times, at intervals of one hour. This solution should be made up fresh just before use or stored in the refrigerator in tightly capped bottles. When bleaching is completed, dehydrate further by passing through 95% and 100% alcohol (30 minutes each). Clear for 30 minutes in a benzol-absolute alcohol mixture (1:1), then transfer to benzyl benzoate for final clearing and permanent storage.

3. The Preparation of Whole Mounts of Fetuses and Embryos

A number of quite satisfactory methods exist for preparation of whole mounts; they may be found in most of the manuals of histology. A point to be noted is that fetuses and implanted embryos are quite fragile and require extremely careful handling to avoid tearing and even dismemberment during preparation. It is very worthwhile to formalin-fix intact uterine segments containing the conceptus before attempting to remove the embryo. Fixative should be continued at least overnight and preferably for 24 hours or longer; formalin acts as a hardening agent, and prolonged exposure to it greatly reduces the risk of damage. Following primary fixation, the formalin may be replaced by 70% alcohol to facilitate dissection of the implantation site and subsequent handling of the embryo. The method given below (from Porturas) is suitable for embryos of nine days ovulation age or older.

Whole mounts of fetuses and late embryos (nine days or older) may be prepared as follows. Uterine horns are removed, pinned out, and fixed in 10% formalin. After a few hours, transect the horns near the implantation sites in order to allow good penetration of the fixative. Continue fixation overnight. Transfer the embryo-containing segment to tap water and dissect out the entire deciduoma with watchmaker's forceps; the method for doing this and for subsequent removal of intact embryos is given on p. 72, and illustrated in Figure 31. Although this method works well with unfixed embryos, these are extremely delicate and success is less certain. However, if several embryos are available, they may be removed within the yolk sac as described, and the yolk sac torn open under formalin. In any case, the embryos should fix at least overnight.

Well-fixed embryos are washed in tap water for several changes of about 10 minutes each and stained in 0.25% Harris' hematoxylin for 15 minutes. They are then transferred to lithium carbonate solution, followed by two changes of tap water. Dehydrate by ascending through alcohols (25%, 50%, 70%, 95% and 100%) in 30-minute steps, repeating the change in 100% alcohol at least once. The dehydrated embryos may then be cleared in cedarwood oil, which may require several changes of a few hours each. Using well slides, permanent mounts may be made using a mounting medium such as Permount.

Other stains, such as carmine, may be used, and different clearing agents may be preferred as well. Texts of histological methods should be consulted for alternative methods of preparing whole mounts.

4. Electron Microscopy of Ova

Enders and Schlafke (32) have studied the ultrastructure of mouse ova in detail; the reference cited describes the methods used. These workers obtained excellent fixation by using cold 3% glutaraldehyde in phosphate buffer for one hour or more. For staining membranes and organelles, ova were postfixed for a similar length of time in phosphate-buffered 2% Osmium tetroxide. Fixed ova were dehydrated in ethyl alcohols, cleared in propylene oxide, and embedded in Araldite epoxy resin. Sections were stained with lead citrate by standard procedures. Fixation in potassium permanganate also gave good results.

R. THE CHEMICAL ANALYSIS
OF OVA

Direct constitutional analysis of mammalian ova has progressed slowly due to the small size of ova and the consequent need to obtain large numbers for most analyses. The wet weight of a mouse ovum is approximately 2.0 μg., about half of which is zona. On the other hand, hormonal superovulation permits the collection of hundreds, and even thousands, of ova without great trouble or expense. In addition, better methods are being evolved for the detection of compounds in small samples. Within recent years, considerable progress has been made in direct and indirect analysis of the composition of mammalian ova. This work is especially promising, in part because of the evidence for precocious determination in the mammalian ovum and the consequent prospect that some elements (enzymes in particular) may appear earlier than expected from developmental studies upon other vertebrate and invertebrate embryos.

Because the methods used for compositional analysis of ova are varied, no attempt will be made to detail them here. Reports concerning composition of ova, or utilization or production of ova, most of them quite recent, deal with CO_2 (19), protein (20, 59, 103), amino acid incorporation (41, 44, 107), sugars (8, 16-18, 104-6, 111), lipid (59), nucleic acid (69, 70, 74), enzymes (1, 3, 68, 73, 85), sialic acid (94), hormones (75, 109), and glycogen (95). Electrophoretic determinations, using thin-layer chromatography and small inocula at the origin, are coming into use, and liquid scintillation counting of radioactive constituents seems on the verge of feasibility. The near future may hold much for these approaches to mammalian development.

Appendix

A. BASIC MEDIUM FOR OVUM CULTURE (BIGGERS, WHITTEN, AND WHITTINGHAM, 7).

Components are given in grams/liter.

NaCl	5.540	Na pyruvate	0.028
KCl	0.356	Na lactate	2.416
Ca lactate.$5H_2O$	0.527	Glucose	1.000
KH_2PO_4	0.162	(Crystalline BSA)	1.000
$MgSO_4.7H_2O$	0.294	Penicillin G to	100 U/ml.
$NaHCO_3$	2.106	Streptomycin S to	50 μg/ml.

The crystalline bovine serum albumin (BSA) is included in parentheses since it is usually necessary to prepare an aliquot of the medium without it for purposes of oil equilibration. In mixing components, all but the phosphate, magnesium salt, and bicarbonate are dissolved in about 600 ml. of distilled water, with continuous stirring. The other three components are dissolved separately in approximately 100 ml. volumes of water and added as such to the rapidly stirring mixture; the bicarbonate solution is usually added last. When all components are in solution, adjust the final volume. Sterilize by filtration, discarding the first 50-100 ml. to pass the filter.

B. MULNARD'S OVUM CULTURE MEDIUM (76).

Prepared solutions are mixed as follows:

NaCl (0.154M)	100 ml.	NaH_2PO4 (0.154M)	1 ml.
KCl (0.154M)	4 ml.	Phenol red (0.2 μg/ml.)	1.2 ml.
$CaCl_2$ (0.1M)	3 ml.	$NaHCO_3$ (0.154M)	to pH 7.4
$MgSO_4$ (0.154M)	1 ml.	Mouse serum*	to 25%

*Mouse serum is collected by cardiac puncture from four-day pregnant females, heated to 56°C. for 30 min., and stored in the refrigerator until used. Antibiotics are omitted; media are filter-sterilized.

C. OOCYTE CULTURE MEDIUM
(BIGGERS WHITTINGHAM, AND
DONAHUE, 8).

Components are given in grams/liter.

NaCl	6.960	NaHCO$_3$	2.106
KCl	0.356	Na pyruvate	0.028
CaCl$_2$	0.189	(Crystalline BSA)	1.000
KH$_2$PO$_4$	0.162	Penicillin G to	100 U/ml.
MgSO$_4$.7H$_2$O	0.294	Streptomycin S to	50 μg/ml.

As with the medium of section A, the salts with divalent cations, plus the phosphate and bicarbonate, are dissolved separately and added last. Filter-sterilize as above.

D. MEDIUM FOR FUSION
ALLOPHENES (MINTZ, 67)

The basis for this medium is Earle's balanced saline solution; it is usually purchased in sterile solution, as are the other ingredients. They may be mixed in a previously autoclaved bottle.

To make 100 ml., components may be mixed as follows:

Earle's BSS	40 ml.
Fetal bovine serum	50 ml.
Phenol red (0.5% solution)	0.4 ml.
Lactic acid (40% solution)	0.25 ml.
Na bicarbonate (7.5% solution)	adjust to pH 7.0

The sterile ingredients are mixed in order; the bicarbonate solution is added last, while mixing rapidly, until the phenol red color indicates a pH near 7.0. The medium is then adjusted to volume and incubated at 37° C. in 5% CO$_2$, with the cap of the bottle loose to permit CO$_2$ equilibration. After several hours the pH will have dropped. It is finally adjusted by the addition of slightly more of the bicarbonate solution if necessary. For this purpose it is convenient to have on hand a set of phenol red standards. After adjustment of pH, return the loosely stoppered bottle to the incubator for an hour or so for final equilibration, then store at 5° C. with the stopper tightly sealed.

Mintz avoids using antibiotics, thus requiring careful asepsis. However, penicillin and streptomycin (100 U/ml. and 50 μg/ml., respectively) are recommended for those without extensive experience in aseptic technique.

E. STAGES IN DEVELOPMENT
OF THE MOUSE EMBRYO

For some reason, the mouse embryo has never been "staged" in the sense of preparation of a complete series of drawings or photographs which state concisely the ex-

ternally visible features that permit recognition of gestational age by inspection of gross specimens. This is
true in spite of the fact that a number of detailed descriptions exist of mouse embryos at various times of development. But these are scattered and often contain much
more information than needed for simple identification
of stage. The drawings given below, except stages 11 to
13, have been redone after original material but have
been guided by drawings and descriptions of Gates (39),
Gruneberg (42), New (78), Otis and Brent (84), and Rugh
(86). Embryos at stages 11 through 13 (six-and-a-half to
nine days) are embedded within the endometrium and are
small, so that precise recognition from external features
is difficult or impossible. Hence these stages are represented by cross-sectional drawings after those of Rugh
(86) and Snell and Stevens (93).

Age based upon assumed time after ovulation is given
in hours (h) or days (d). Scale next to Stage 2 applies
through Stage 9. S 10: ovary larger than conceptual
swellings; developing blastocyst football shaped; S 11:
swellings about the same size as ovary; S 12: swellings
about 50% larger than ovary; S 13: swellings about twice
as large as ovary; embryo recognizable but inverted;
S 14: hind limb bud not formed; no nasal process, olfac-
tory pit, or optic cup bulge; S 15; nasal processes widely
separated and deep; anterior limb buds spade-like and
well defined, posterior buds less so; S 16: nasal processes
in close contact and form a slit-like nasal pit; auditory
vesicles not visible, pinna not formed; in forelimb, paw
separated from the remainder of limb by a constriction;
no paw separation in hind limb; S 17: naso-maxillary
fissure fused; three rows of whiskers present (not shown),
containing three to four follicles each; anterior foot plate
with rays; posterior plate smooth and circular; S 18: five
rows of whiskers (not shown) with five to seven follicles
each; anterior foot plate with laterally indented digits
outlined; posterior plate with rays only; marked umbilical
hernia; S 19: pinna flap-like and turned forward, covering
about a third of the meatus; forepaw digits separate dis-
tally, webbed at the base; hair follicles on body except
head and mid-back; tail segmentation not visible; S 20:
flap of pinna covers most of meatus; fore and hind limbs
unwebbed, divergent; hair follicles everywhere except
eyelids; S 21: flap of pinna covers meatus; eyelids often
beginning to close; end phalanges distinct in fore and

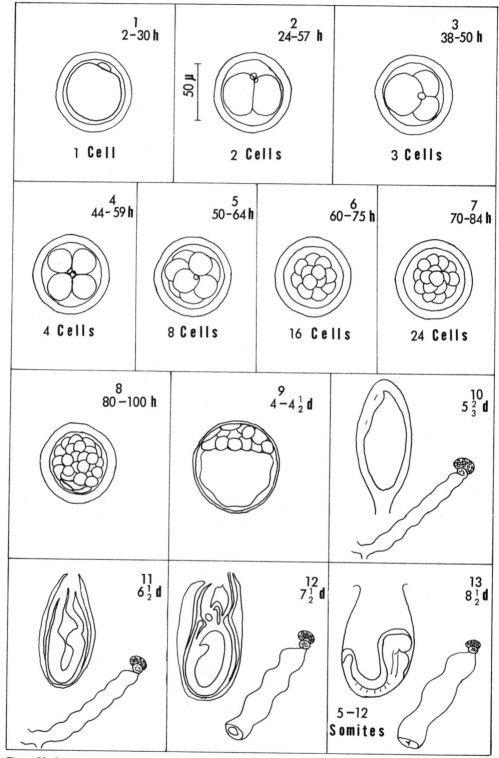

Figure 33. Stages in Development of the Mouse I.

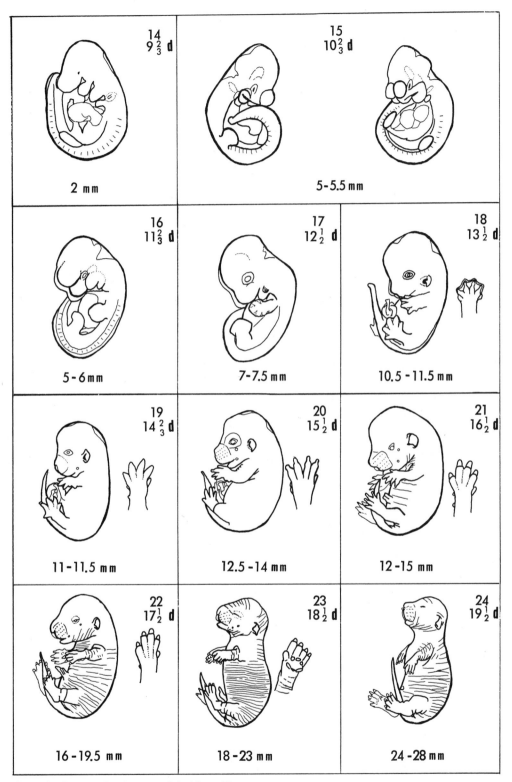

Figure 34. Stages in Development of the Mouse II.

hind digits; umbilical hernia reduced; S 22: eyelids usually found to be closed; digits nearly parallel; wrinkling of skin pronounced; S 23: head skin wrinkled; eyes closed; digits parallel; back arched; S 24: forelimbs dropped, on a level with tip of tail; axis of head parallel with hind-limbs. This is the earliest stage at which parturition nor-mally occurs. Criteria for Stages 14 et seq. after Grune-berg (42).

F. PLANS FOR CONSTRUCTION OF A SUITABLE CLEAN HOOD

Parts may be cut from a sheet of Plexiglas 4 feet x 6 feet, as shown. Material should be ¼ inch thick. The top of the front piece must be beveled for close fit, allowed for in dimensions. Pieces may be screwed together, or clamped in position and fused with chloroform. Long corner pieces (not shown on plans) are desirable for added rigidity. For view of assembled unit, see Figure 1. Plan taken from units in use in J. D. Biggers' laboratory.

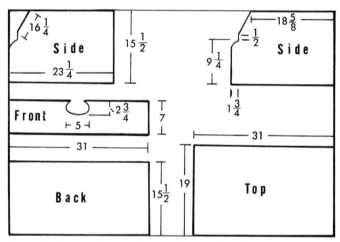

Figure 35. Plans for Construction of a Suitable Clean Hood.

Useful Reference Sources

The following books are especially useful for additional details concerning development and husbandry of the mouse, as well as methodology.

Austin, C. R. 1961. *The Mammalian Egg.* Oxford: Blackwell Scientific Publishers.

Burdette, W. J., ed. 1963. *Methodology in Mammalian Genetics.* San Francisco: Holden-Day, Inc.

Daniel, J. C., ed. In press. *Methods of Mammalian Embryology.* San Francisco and London: W. H. Freeman & Co.

Green, E. L., ed. 1966. *Biology of the Laboratory Mouse.* New York: The Blakiston Division, McGraw-Hill Book Co.

New, D. A. T. 1966. *The Culture of Vertebrate Embryos.* New York and London: Academic Press.

Parker, R. C. 1961. *Methods of Tissue Culture.* New York: Hoeber-Harper.

Rugh, R. 1968. *The Mouse.* Minneapolis: Burgess Publishing Co.

Wilt, F. H., and Wessels, N. K., eds. 1967. *Methods in Developmental Biology.* New York: Thomas Y. Crowell Co.

Wolstenholme, G. E. W., and O'Connor, M., eds. 1965. *Preimplantation Stages of Pregnancy.* CIBA Foundation Symposium. Boston: Little, Brown & Co.

A Bibliography of Selected Papers

1. Auerbach, S., and Brinster, R. L. 1968. Lactate dehydrogenase enzymes in mouse blastocyst cultures. *Exptl. Cell Res.* 53: 313-15.
2. Austin, C. R. 1961. *The Mammalian Egg.* Oxford: Blackwell Scientific Publications.
3. Baker, W. W., and Mintz, B. 1969. Subunit structure and gene control of mouse NADP-malate dehydrogenase. *Biochem. Genetics* 2: 351-60.
4. Biggers, J. D., and Brinster, R. L. 1965. Biometrical problems in the study of early mammalian embryos *in vitro. J. Exp. Zool.* 158: 39-48.
5. Biggers, J. D., Gwatkin, R. B. L., and Brinster, R. L. 1962. Development of mouse embryos in organ culture of fallopian tubes on a chemically defined medium. *Nature* 194: 747-49.
6. Biggers, J. D., Moore, B. D., and Whittingham, D. G. 1965. Development of mouse embryos *in vivo* after cultivation from two-cell ova to blastocysts *in vitro. Nature* 206: 734-35.
7. Biggers, J. D., Whitten, W. K., and Whittingham, D. G. In press. The culture of mouse embryos *in vitro.* In *Methods of mammalian embryology,* ed. J. C. Daniel. San Francisco and London: W. H. Freeman & Co.
8. Biggers, J. D., Whittingham, D. G., and Donahue, R. P. 1967. The pattern of energy metabolism in the mouse oocyte and zygote. *Proc. Natl. Acad. Sci.* 58: 560-67.
9. Billington, W. D. 1964. Extrauterine development of mouse trophoblast. *J. Reprod. Fertil.* 8: 274-75.
10. Billington, W. D., Graham, C. F., and McLaren, A. 1968. Extra-uterine development of mouse blastocysts cultured *in vitro* from early cleavage stages. *J. Embryol. Exptl. Morph.* 20: 391-400.
11. Brackett, B. G., and Williams, W. L. 1965. *In vitro* fertilization of rabbit ova. *J. Exp. Zool.* 160: 271-82.
12. _____. 1968. Fertilization of rabbit ova in a defined medium. *Fertil. Steril.* 19: 144-56.
13. Braden, A. W. H., and Austin, C. R. 1954. The fertile life of mouse and rat eggs. *Science* 120: 610-11.
14. Brinster, R. L. 1963. A method for *in vitro* cultivation of mouse ova from two-cell to blastocyst. *Exptl. Cell Res.* 32: 205-08.
15. _____. 1965. Studies on the development of mouse embryos *in vitro* I: The effect of osmolarity and hydrogen ion concentration. *J. Exp. Zool.* 158: 49-58.

16. _____. 1965. Studies on the development of mouse embryos *in vitro* II: The effect of energy source. *J. Exp. Zool.* 158: 59-68.

17. _____. 1965. Studies on the development of mouse embryos *in vitro* III: The effect of fixed nitrogen source. *J. Exp. Zool.* 158: 69-77.

18. _____. 1965. Studies on the development of mouse embryos *in vitro* IV: Interaction of energy sources. *J. Reprod. Fertil.* 10: 227-40.

19. _____. 1967. Carbon dioxide production from glucose by the preimplantation mouse embryo. *Exptl. Cell Res.* 47: 271-77.

20. _____. 1967. Protein content of the mouse embryo during the first five days of development. *J. Reprod. Fertil.* 13: 413-20.

21. _____. 1969. Mammalian embryo culture. In *The Mammalian Oviduct,* eds. E. S. E. Hafez and R. J. Blandau. Chicago: The University of Chicago Press, pp. 419-44.

22. Brinster, R. L., and Biggers, J. D. 1965. *In vitro* fertilization of mouse ova within the explanted fallopian tube. *J. Reprod. Fertil.* 10: 277-79.

23. Bronson, F. H., Dagg, C. P., and Snell, G. D. 1966. Reproduction. In *Biology of the Laboratory Mouse,* ed. E. L. Green. New York: McGraw-Hill Book Co., pp. 187-204.

24. Chang, M. C. 1955. The maturation of rabbit oocytes in culture and their maturation, activation, fertilization and subsequent development in the fallopian tubes. *J. Exp. Zool.* 128: 379-410.

25. _____. 1959. Fertilization of rabbit ova *in vitro*. *Nature* 184: 466-67.

26. Deringer, M. K. 1963. Technique for the transfer of fertilized ova. In *Methodology in Mammalian Genetics,* ed. W. J. Burdette. San Francisco: Holden-Day, Inc., pp. 563-64.

27. Dickie, M. M. 1963. Methods of keeping records. In *Methodology in Mammalian Genetics,* ed. W. J. Burdette. San Francisco: Holden-Day, Inc., pp. 522-37.

28. Donahue, R. P. 1968. Maturation of the mouse egg *in vitro*. Ph.D. Thesis, the Johns Hopkins University School of Medicine.

29. Donahue, R. P., and Stern, S. 1968. Follicular cell support of oocyte maturation: Production of pyruvate *in vitro*. *J. Reprod. Fertil.* 17: 395-98.

30. Dziuk, P. J., and Runner, M. N. 1960. Recovery of blastocysts and induction of implantation following artificial insemination of immature mice. *J. Reprod. Fertil.* 1: 321-31.

31. Edwards, R. G. 1965. Maturation *in vitro* of mouse, sheep, cow, pig, Rhesus monkey and human ovarian oocytes. *Nature* 208: 349-51.

32. Enders, A. C., and Schlafke, S. J. 1965. The fine structure of the blastocyst: Some comparative studies in pre-implantation stages. In *Preimplantation Stages of Pregnancy,* CIBA Foundation Symposium, eds. G. E. W. Wolstenholme and M. O'Connor. Boston: Little, Brown & Co., pp. 29-54.

33. Fawcett, D. W. 1950. The development of mouse ova under the capsule of the kidney. *Anat. Rec.* 108: 71-92.

34. Fawcett, D. W., Wislocki, G. B., and Waldo, C. M. 1947. The development of mouse ova in the anterior chamber of the eye and in the abdominal cavity. *Am. J. Anat.* 81: 413-44.

35. Fekete, E., and Little, C. C. 1942. Observations on the mammary tumor incidence in mice born from tranferred ova. *Cancer Res.* 2: 525-30.

36. Fowler, R. E., and Edwards, R. G. 1957. Induction of superovulation and pregnancy in mature mice by gonadotrophins. *J. Endocrinol.* 15: 374-84.

37. Gardner, R. L. 1968. Mouse chimaeras obtained by the injection of cells into the blastocyst. *Nature* 220: 596-97.

38. Gates, A. 1956. Viability and developmental capacity of eggs from immature mice treated with gonadotrophins. *Nature* 177: 754-55.

39. ———. 1965. Rate of ovular development as a factor in embryonic survival. In *Preimplantation Stages of Pregnancy,* CIBA Foundation Symposium, eds. G. E. W. Wolstenholme and M. O'Connor. Boston: Little, Brown & Co., pp. 270-88.

40. Glenister, T. W. 1963. Observations on mammalian blastocysts implanting in organ culture. In *Delayed Implantation,* ed. A. Enders. Houston: *Rice University Semicentennial Publications,* pp. 171-82.

41. Greenwald, G. S., and Everett, N. B. 1959. The incorporation of S^{35} methionine by the uterus and ova of the mouse. *Anat. Rec.* 134: 171-84.

42. Gruneberg, H. 1943. The development of some external features in mouse embryos. *J. Hered.* 34: 88-92.

43. Gwatkin, R. B. L. 1964. Effect of enzymes and acidity on the zona pellucida of the mouse egg before and after fertilization. *J. Reprod. Fertil.* 7: 99-105.

44. ———. 1966. Amino acid requirements for attachment and outgrowth of the mouse blastocyst *in vitro. J. Cell Physiol.* 68: 335-43.

45. ———. 1966. Defined media and development of mammalian eggs *in vitro. Ann. N.Y. Acad. Sci.* 139: 79-90.

46. Hammond, J., Jr. 1949. Recovery and culture of tubal mouse ova. *Nature* 163: 28-9.

47. Hoag, W. M., and Les, E. P. 1963. Husbandry, equipment and procurement of mice. In *Methodology in Mammalian Genetics,* ed. W. J. Burdette. San Francisco: Holden-Day, Inc., pp. 538-57.

48. Jensen, F. C., Gwatkin, R. B. L., and Biggers, J. D. 1964. A simple organ culture method which allows simultaneous isolation of specific types of cells. *Exptl. Cell Res.* 34: 440-47.

49. Kallman, R. F. 1967. The mouse. In *Methods in Developmental Biology,* eds. F. H. Wilt and N. K. Wessels. New York: Thomas Y. Crowell Co., pp. 3-12.

50. Kennedy, J. F., and Donahue, R. P. 1969. Human oocytes: Maturation in chemically defined medium. *Science* 164: 1292-3.

51. Kirby, D. R. S. 1960. Development of mouse eggs beneath the kidney capsule. *Nature* 187: 707-8.

52. ———. 1962. The influence of the uterine environment on the development of mouse eggs. *J. Embryol. Exp. Morph.* 10: 465-506.

53. ———. 1962. Reciprocal transplant of blastocysts between rats and mice. *Nature* 194: 785-86.

54. ———. 1963. The development of mouse blastocysts transplanted to the scrotal and cryptorchid testis. *J. Anat.* (London), 97: 119-30.

55. ———. 1963. Development of the mouse blastocyst transplanted to the spleen. *J. Reprod. Fertil.* 5: 1-12.

56. Krzanowski, H. 1964. Time interval between copulation and fertilization in inbred lines of mice and their crosses. *Folia Biologica* 12: 231-44.

57. Lewis, W. H., and Wright, E. S. 1935. On the early development of the mouse egg. *Carnegie Contributions to Embryology* 25: 115-44.

58. Lin, T. P. 1966. Microinjection of mouse eggs. *Science* 151: 333-37.

59. Loewenstein, J. E., and Cohen, A. I. 1964. Dry mass, lipid content and protein content of the intact and zona free mouse ovum. *J. Embryol. Expt. Morph.* 12: 113-19.

60. Marston, J. H., and Chang, M. C. 1964. The fertilizable life of ova and their morphology following delayed insemination in mature and immature mice. *J. Exp. Zool.* 155: 237-52.

61. McLaren, A., and Biggers, J. D. 1958. Successful development and birth of mice cultivated *in vitro* as early embryos. *Nature* 182: 877-78.

62. McLaren, A., and Michie, D. 1956. Studies on the transfer of fertilized mouse eggs to uterine foster-mothers I: Factors affecting the implantation and survival of native and transferred eggs. *J. Exptl. Biol.* 33: 394-416.

63. _____. 1959. Studies on the transfer of fertilized mouse eggs to uterine foster mothers II: The effect of transferring large numbers of eggs. *J. Exptl. Biol.* 36: 40-50.

64. McLaren, A., and Tarkowski, A. K. 1963. Implantation of mouse eggs in the peritoneal cavity. *J. Reprod. Fertil.* 6: 384-92.

65. Mills, R. M., and Brinster, R. L. 1967. Oxygen consumption of preimplantation mouse embryos. *Exptl. Cell Res.* 47: 337-44.

66. Mintz, B. 1962. Experimental study of the developing mammalian egg: removal of the zona pellucida. *Science* 138: 594-95.

67. _____. 1964. Formation of genetically mosaic mouse embryos and early development of "lethal (t^{12}/t^{12})-normal" mosaics. *J. Exp. Zool.* 157: 273-92.

68. _____. 1964. Synthetic processes and early development in the mammalian egg. *J. Exp. Zool.* 157: 85-100, 267-94.

69. _____. 1964. Gene expression in the morula stage of mouse embryos, as observed during development of t^{12}/t^{12} lethal mutants *in vitro*. *J. Exp. Zool.* 157: 267-72.

70. _____. 1965. Nucleic acid and protein synthesis in the developing mouse embryo. In *Preimplantation Stages of Pregnancy,* CIBA Foundation Symposium, eds. G. E. W. Wolstenholme and M. O'Connor. Boston: Little, Brown & Co., pp. 145-55.

71. _____. 1967. Gene control of mammalian pigment differentiation I: Clonal origin of melanocytes. *Proc. Natl. Acad. Sci.* 58: 344-51.

72. _____. 1967. Mammalian embryo culture. In *Methods in developmental biology,* eds. F. H. Wilt and N. K. Wessels. New York: Thomas Y. Crowell Co., pp. 379-400.

73. Mintz, B., and Baker, J. 1967. Normal mammalian muscle differentiation and gene control of isocitrate dehydrogenase synthesis. *Proc. Natl. Acad. Sci.* 58: 502-98.

74. Monessi, V., and Salfi, V. 1967. Macromolecular synthesis during early development in the mouse embryo. *Exptl. Cell Res.* 46: 632-35.

75. Moore-Smith, D. 1968. Implantation in the mouse. The effect on implantation of treating cultured mouse blastocysts with estrogen *in vitro,* and blastocyst uptake of H^3-oestradiol. *J. Endocrinol.* 41: 17-29.

76. Mulnard, J. G. 1964. Obtentien in vitro du développement continu de l'oeuf de souris du stade II au stade du blastocyste. *C. R. Acad. Sci.* (Paris) 258: 6228-29.

77. _____. 1965. Studies of regulation of mouse ova *in vitro*. In *Preimplantation Stages of Pregnancy,* CIBA Foundation Symposium, eds. G. E. W. Wolstenholme and M. O'Connor. Boston: Little, Brown & Co., pp. 123-38.

78. New, D. A. T. 1966. *The Culture of Vertebrate Embryos.* New York and London: Academic Press, pp. 20-46.

79. _____. 1966. Development of rat embryos cultured in blood sera. *J. Reprod. Fertil.* 12: 509-24.

80. New, D. A. T., and Stein, K. F. 1963. Cultivation of mouse embryos *in vitro*. *Nature* 199: 297-99.

81. _____. 1964. Cultivation of post-implantation mouse and rat embryos on plasma clots. *J. Embryol. Exp. Morph.* 12: 101-11.

82. Noyes, R. W., and Dickman, Z. 1961. Survival of ova transferred into the oviduct of the rat. *Fertil. Steril.* 12: 67-79.

83. Orsini, M. W. 1962. Technique of preparation, study and photography of benzyl-benzoate material for embryological studies. *J. Reprod. Fertil.* 3: 283-87.

84. Otis, E. M., and Brent, R. 1954. Equivalent ages in mouse and human embryos. *Anat. Rec.* 120: 33-63.

85. Rapola, J., and Koskimies, O. 1967. Embryonic enzyme patterns: Characterization of the single lactate dehydrogenase isozyme in preimplanted mouse ova. *Science* 157: 1311-12.

86. Rugh, R. 1968. *The Mouse, Its Reproduction and Development.* Minneapolis: Burgess Publishing Co.

87. Runner, M. N. 1947. Development of mouse eggs in the anterior chamber of the eye. *Anat. Rec.* 98: 1-14.

88. _____. 1951. Differentiation of intrinsic and maternal factors governing intrauterine survival of mammalian young. *J. Exp. Zool.* 116: 1-20.

89. Runner, M. N., and Gates, A. 1954. Sterile, obese mothers. *J. Hered.* 45: 51-5.

90. Runner, M. N. and Palm, J. 1953. Transplantation and survival of unfertilized ova of the mouse in relation to postovulatory age. *J. Exp. Zool.* 124: 303-16.

91. Seidel, F. 1960. Die Entwicklungsfähigkeiten isolierter Furchungszellen aus dem Ei des Kaninchens *Oryctolagus cuniculus. Wilhelm Roux' Arch. EntMech. Org.* 152: 43-130.

92. Smith, L. J. 1964. The effects of transection and extirpation on axis formation and elongation in the young mouse embryo. *J. Embryol. Exp. Morph.* 12: 787-803.

93. Snell, G. D., and Stevens, L. C. 1966. Early embryology. In *Biology of the Laboratory Mouse,* ed. E. L. Green. New York: McGraw-Hill, pp. 205-45.

94. Soupart, P., and Noyes, R. W. 1964. Sialic acid as a component of the zona pellucida of the mammalian ovum. *J. Reprod. Fertil.* 8: 251-53.

95. Stern, S., and Biggers, J. D. 1968. Enzymatic estimation of glycogen in the cleaving mouse embryo. *J. Exp. Zool.* 168: 61-66.

96. Tarkowski, A. K. 1959. Experiments on the transplantation of ova in mice. *Acta Theriol.* 2: 251.

97. _____. 1959. Experimental studies on regulation in the development of isolated blastomeres of mouse eggs. *Acta. Theriol.* 3: 191-267.

98. _____. 1961. Foster transfers of mouse chimaeras developed from fused eggs. *Nature* 190: 857-60.

99. _____. 1961. Mouse chimaeras developed from fused eggs. *Nature* 190: 857-60.

100. _____. 1963. Studies on mouse chimeras developed from eggs fused *in vitro. Natl. Cancer Inst. Monograph* 11: 51-71.

101. _____. 1965. Embryonic and postnatal development of mouse chimeras. In *Preimplantation Stages of Pregnancy,* CIBA Foundation Symposium, eds. G. E. W. Wolstenholme and M. O'Connor. Boston: Little, Brown & Co., pp. 183-207.

102. Tarkowski, A. K., and Wroblewska, J. 1967. Development of blastomeres of mouse eggs isolated at the 4- and 8-cell stage. *J. Embryol. Exptl. Morph.* 18: 155-80.

103. Thompson, J. L., and Biggers, J. D. 1966. Effect of inhibitors of protein synthesis on the development of preimplantation mouse embryos. *Exptl. Cell Res.* 41: 411-27.

104. Wales, R. G., and Biggers, J. D. 1968. The permeability of two- and eight-cell mouse embryos to L-malic acid. *J. Reprod. Fertil.* 15: 103-12.

105. Wales, R. G., and Brinster, R. L. 1968. The uptake of hexoses by preimplantation mouse embryos *in vitro. J. Reprod. Fertil.* 15: 514-522.

106. Wales, R. G. and Whittingham, D. G. 1967. A comparison of the uptake and utiliza-

tion of lactate and pyruvate by one- and two-cell mouse embryos. *Biochim. Biophys. Acta* 148: 703-12.

107. Weitlauf, H. M., and Greenwald, G. S. 1967. A comparison of the *in vivo* incorporation of S^{35} methionine by two-celled mouse eggs and blastocysts. *Anat. Rec.* 159: 249-54.

108. Whitten, W. K. 1956. Culture of tubal mouse ova. *Nature* 177: 96.

109. _____. 1957. The effect of progesterone on the development of mouse eggs *in vitro*. *J. Endocrinol.* 16: 80-5.

110. _____. 1966. Pheromones and mammalian reproduction. *Adv. Reprod. Physiol.* 1: 155-78.

111. Whitten, W. K., and Biggers, J. D. 1968. Complete culture of the preimplantation stages of the mouse *in vitro*. *J. Reprod. Fertil.* 17: 399-402.

112. Whitten, W. K., and Dagg, C. P. 1961. Influence of spermatozoa on the cleavage rate of mouse eggs. *J. Exp. Zool.* 148: 173-83.

113. Whittingham, D. G. 1967. Studies on the early preimplantation stages of mammalian development. Ph.D. Thesis, University of London.

114. _____. 1968. Fertilization of mouse eggs *in vitro*. *Nature* 220: 592-93.

115. _____. 1968. Development of zygotes in cultured mouse oviducts I: The effect of varying oviducal conditions. *J. Exp. Zool.* 169: 391-97.

116. Whittingham, D. G., and Biggers, J. D. 1967. Fallopian tube and early cleavage in the mouse. *Nature* 213: 942-43.

117. Yanagimachi, R., and Chang, M. C. 1963. Fertilization of hamster eggs *in vitro*. *Nature* 200: 281-82.

Index

Numbers printed in boldface type indicate pages where illustrations occur.

THE JOHNS HOPKINS PRESS

Designed by Arlene J. Sheer

*Composed in Times Roman text and display
by University Graphics, Inc.*

*Printed on Finch Text Book for Offset
by Universal Lithographers, Inc.*

*Bound in Columbia Riverside
by Moore and Company, Inc.*